Coherent Time Difference of Arrival Estimation Techniques for Frequency Hopping GSM Mobile Radio Signals

von

Dr.-Ing. Alexander Gerald Götz
Friedrich-Alexander-University, Erlangen-Nuremberg

Oldenbourg Verlag München

Dr.-Ing. Alexander Gerald Götz studied Electrical Engineering, Electronics and Information Technology at Friedrich-Alexander-University Erlangen-Nuremberg, Germany and graduated as Dipl.-Ing. Univ. in May 2008. Subsequently he was engaged as research assistant and doctoral candidate at the Institute for Electronics Engineering (Prof. R. Weigel) at Friedrich-Alexander-University Erlangen-Nuremberg, Germany and graduated as Dr.-Ing. in September 2012. At the Institute for Electronics Engineering he conducted research in the area of GSM mobile phone localization in search and rescue scenarios. His main research interests are in the field of communication engineering, localization and radar technology, digital signal processing, radio frequency engineering as well as analog and digital circuit design.

Bibliografische Information der Deutschen Nationalbibliothek

Die Deutsche Nationalbibliothek verzeichnet diese Publikation in der Deutschen Nationalbibliografie; detaillierte bibliografische Daten sind im Internet über http://dnb.d-nb.de abrufbar.

© 2013 Oldenbourg Wissenschaftsverlag GmbH
Rosenheimer Straße 143, D-81671 München
Telefon: (089) 45051-0
www.oldenbourg-verlag.de

Lektorat: Johannes Breimeier
Herstellung: Constanze Müller
Einbandgestaltung: hauser lacour
Gesamtherstellung: Books on Demand GmbH, Norderstedt

Dieses Papier ist alterungsbeständig nach DIN/ISO 9706.

ISBN   978-3-486-73178-1
eISBN   978-3-486-74862-8

Coherent Time Difference of Arrival Estimation Techniques for Frequency Hopping GSM Mobile Radio Signals

Kohärente Verfahren zur Laufzeitdifferenzschätzung für frequenzspringende GSM-Mobilfunksignale

Der Technischen Fakultät der
Universität Erlangen-Nürnberg
zur Erlangung des Grades

DOKTOR-INGENIEUR

vorgelegt von

Dipl.-Ing. Univ.
Alexander Gerald Götz

Erlangen - 2012

Als Dissertation genehmigt von
der Technischen Fakultät der
Universität Erlangen-Nürnberg

Tag der Einreichung: 8. Juni 2012
Tag der Promotion: 17. September 2012

Dekanin der Fakultät: Prof. Dr.-Ing. habil. Marion Merklein
1. Berichterstatter: Prof. Dr.-Ing. Dr.-Ing. habil. Robert Weigel
2. Berichterstatter: Prof. Dr.-Ing. Martin Vossiek

Abstract

In this work, coherent techniques for time difference of arrival estimation for frequency hopping GSM signals are introduced. The techniques provide significant improvements in accuracy compared to state-of-the-art techniques and are ideally suited for highly accurate localization of GSM mobile phones.

The key inventive concept is based on the interpretation of a frequency hopping GSM signal as a wideband signal. Thus, the applicable bandwidth for time difference of arrival estimation can be increased from $200\,\mathrm{kHz}$ for the narrowband burst signal to the full uplink bandwidth of the corresponding GSM standard. For E-GSM 900 systems, up to $35\,\mathrm{MHz}$ of bandwidth can be employed. Consequently, a localization accuracy in the scale of $5 - 10\,\mathrm{m}$ is achievable.

An essential requirement of the presented techniques is the phase preserving, i.e. coherent acquisition and processing of the signals. For coherent signal acquisition, two possible frontend implementations are proposed and the corresponding signals are modeled. Furthermore, two different coherent techniques for time difference of arrival estimation are analytically derived and investigated.

The performance of the coherent techniques is compared to a representative incoherent state-of-the-art technique by computer simulations. The main emphasis of the investigations is on the noise performance and multipath resolvability of the techniques. The performance of the presented coherent techniques is shown to be superior compared to the incoherent state-of-the-art technique.

The applicability of the presented techniques in real scenarios is verified by a hardware prototype system. The prototype system architecture shows to meet the fundamental estimation requirements and the measurement results provide a proof-of-concept. Further research on this topic is therefore encouraged.

The presented coherent techniques for time difference of arrival estimation permit novel applications with increased accuracy requirements such as highly accurate localization in search and rescue scenarios. Furthermore, the coherent estimation concept can also be adapted to any frequency hopping signal source such as TETRA, DECT, IEEE 802.15.4 (ZigBee) and IEEE 802.15.1 (Bluetooth) devices.

Zusammenfassung

Die vorliegende Arbeit befasst sich mit Verfahren zur Laufzeitdifferenzschätzung von frequenzspringenden GSM-Mobilfunksignalen. Die vorgestellten Verfahren ermöglichen wesentliche Verbesserungen der Schätzgenauigkeit im Vergleich zum Stand der Technik und sind für die hochgenaue Ortung von GSM-Mobiltelefonen ausgelegt.

Die Kernidee besteht in der Interpretation des frequenzspringenden GSM-Mobilfunksignals als Breitbandsignal. Auf diese Weise kann die nutzbare Bandbreite zur Laufzeitdifferenzschätzung von $200\,\text{kHz}$ für ein schmalbandiges Burstsignal bis auf die volle Bandbreite der Aufwärtsstrecke des zugehörigen GSM-Standards erweitert werden. Für E-GSM 900 stehen somit bis zu $35\,\text{MHz}$ an Bandbreite zur Verfügung. Somit wird eine Ortungsgenauigkeit in der Größenordnung von $5 - 10\,\text{m}$ ermöglicht.

Eine wesentliche Voraussetzung für die vorgestellten Verfahren ist die phasenerhaltende, d.h. kohärente Erfassung und Verarbeitung der Signale. Für die kohärente Signalerfassung werden zwei mögliche Implementierungen der Empfängerkomponenten vorgeschlagen und die zugehörigen Empfangssignale modelliert. Des Weiteren werden zwei unterschiedliche kohärente Verfahren zur Laufzeitdifferenzschätzung analytisch hergeleitet und untersucht.

Die Eigenschaften der kohärenten Verfahren werden mittels Rechnersimulationen mit einem repräsentativen inkohärenten Verfahren verglichen, das den gegenwärtigen Stand der Technik darstellt. Der Schwerpunkt der Untersuchungen liegt hierbei im Rauschverhalten sowie der Auflösbarkeit von Mehrwegepfaden. Es wird hierbei gezeigt, dass die vorgestellten kohärenten Verfahren wesentlich bessere Ergebnisse erzielen als das inkohärente Verfahren.

Die Anwendbarkeit der vorgestellten Verfahren in realen Umgebungen wird mittels eines Prototyps verifiziert. Es wird gezeigt, dass die Systemarchitektur des Prototyps die wesentlichen Voraussetzungen für die Laufzeitdifferenzschätzung erfüllt. Zudem bestätigen die Messergebnisse die Anwendbarkeit des vorgestellten Konzepts. Eine weiterführende Forschung in diesem Gebiet erscheint hierdurch vielversprechend.

Die vorgestellten Verfahren zur Laufzeitdifferenzschätzung ermöglichen neuartige Anwendungen mit erhöhten Genauigkeitsanforderungen wie beispielsweise eine hochgenaue Ortung von Verschütteten in Rettungs- und Bergungsszenarien. Des Weiteren kann das vorgestellte Konzept auf beliebige andere Signalquellen mit Frequenzsprungverfahren wie beispielsweise TETRA, DECT, IEEE 802.15.4 (ZigBee) sowie IEEE 802.15.1 (Bluetooth) erweitert werden.

Acknowledgments

The completion of this work would never have been possible without the support of numerous colleagues, friends and family members.

First of all, I would like to express my sincere thanks to Prof. Robert Weigel for the great opportunity to work on this project and the excellent research conditions at the institute. I always appreciated the supportive atmosphere, the open door at any time as well as the possibilities to present my work at conferences worldwide. The confidence in my work and the possibility to pursue a patent application has also been very motivating.

Many thanks for reviewing this work as well as participating in the doctorate committee are addressed to Prof. Martin Vossiek, Prof. Robert Schober and Prof. Marc Stamminger. Their support and engagement for pursuing my doctorate is gratefully acknowledged.

During my time at the institute, numerous colleagues and friends have formed my everyday life and I would like to thank everyone for this exhilarating time. Many thanks go to my project mates Stefan Zorn and Richard Rose for the enjoyable time at field trials and conferences as well as my room mate Anna Gabiger-Rose for the pleasant common time. Furthermore, I would like to express my thanks to Benjamin Waldmann for the great time during scientific and non-scientific events. Many further colleagues have contributed to a worthwhile time at the institute with common activities and interests. Many evenings have created friendships that will last over time.

From the side of industry, I am very grateful for the excellent support from Symeo GmbH, Munich. In particular, I would like to acknowledge the advice and discussions with Peter Gulden throughout the research project.

Since no enduring challenge can be tackled without a supporting personal environment, the contributions of my parents, my brother and my girlfriend can not be overestimated. Their sustaining interest in the progress of my thesis as well as their encouragement in times of setback have been very important for me and are sincerely acknowledged.

This work is dedicated to all the people who have accompanied me during the last four years in Erlangen. All the people who have made my life interesting and enjoyable and have contributed to a worthwhile time at the institute. I am very grateful for this unique time which has shaped my life and will open new perspectives for years to come...

Alexander Gerald Götz
Erlangen, June 2012

Science is organized knowledge. Wisdom is organized life.
(Immanuel Kant, 1724–1804 A.D.)

Contents

List of Figures

List of Tables

1 Introduction and Motivation

In the following, an introduction to the scientific background and a motivation for conducting research in this area are presented. Furthermore, the key innovative concept and the structure of this work is outlined.

1.1 GSM Mobile Phone Localization

The *Global System for Mobile Communications (GSM)* is one of the most successful standards for mobile communications worldwide and the localization of GSM mobile phones has been subject of scientific research over the past few years. Accurate localization has mainly been motivated by the emergence of *Localization Based Services (LBS)* such as location based billing of network subscribers, monitoring and effective dispatch of vehicle fleets and providing location-sensitive information to network subscribers. For network operators, the monitoring of mobile phone positions has been a desirable objective for efficient and effective planning of the cellular network as well as capacity improvements. Another driving factor for mobile phone localization has been the requirement of providing location information in case of emergency calls mandated by the European Commission and the United States Federal Communications Commission known as the E112 and E911 initiatives. [1, 2, 3]

The main objective of mobile phone localization is illustrated in Fig. 1.1.

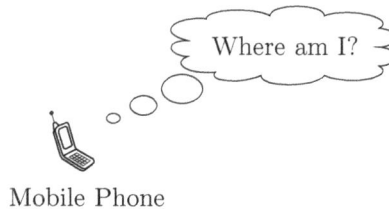

Where am I?

Mobile Phone

Figure 1.1: The main objective of mobile phone localization

The applied techniques for localization in these scenarios have primarily been designed to fit in the existing architecture of mobile phone networks. Within these networks, different techniques for localization have been standardized. The simplified structure of a GSM network cell is depicted in Fig. 1.2.

Base Station B

Base Station A   Base Station C

Mobile Phone

Base Station E   Base Station D

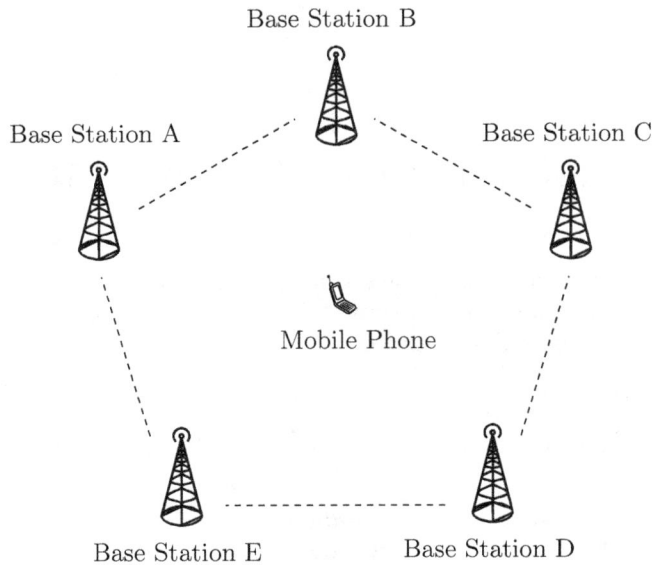

Figure 1.2: Structure of a GSM network cell

The standardized techniques comprise the *Cell Identification (Cell-ID)* method, the sectorization of the cell and evaluation of the *Timing Advance (TA)* value, the evaluation of the *Received Signal Strength (RSS or RXLEV)* value and time-of-flight based approaches such as the *Uplink Time Difference of Arrival (U-TDOA)* and *Enhanced Observed Time Difference (E-OTD)* methods. These techniques require additional *Location Measurement Units (LMUs)* to be installed at the corresponding base station sites. [4]

The cell identification approach is based on the identification number of the serving base station. Depending on this information, an estimate of the mobile phone location in the scale of the cell size is possible. Since the size of the cell varies between 500 m in urban areas and 35 km in rural areas, only a coarse localization is possible. [5]

An improved location estimate can be obtained by sectorization of the cell and evaluation of the timing advance value. The sectorization of the cell provides a directional segment for the mobile phone location. The timing advance value represents a timing offset provided by the serving base station in order to adjust the transmit time of the mobile phone to the corresponding timing of the base station. The timing advance value is characterized by a step size corresponding to 554 m. Therefore, the accuracy is still in the scale of several hundreds of meters. [5]

The evaluation of the received signal strength is another indicator for the distance between the mobile phone and the serving base station. This value is inherently available for power control and can be used for distance estimation. Since the radio channel is characterized by fading and shadowing effects, the accuracy of this approach is also in the scale of several hundreds of meters. A localization method based on this approach is presented in [6]. A combined evaluation of the cell identification, the timing advance value and the received signal strength is investigated in [7].

The uplink time difference of arrival method is based on measuring the time difference of signal reception of a mobile phone signal at three or more base stations. Since the signal originates at the mobile station, the uplink signal band is used for this technique. The enhanced observed time difference method is based on measuring the time differences of signal reception of synchronously transmitted signals from the base stations at the mobile phone. This approach requires software modifications at the mobile phone and can be characterized as inverse approach with respect to the uplink time difference of arrival method. [5, 8]

The main challenge of time difference of arrival based localization approaches is the narrow bandwidth of GSM signals of approximately 200 kHz. Since the resolvability of multipath components is directly related to the bandwidth of the applied signals, the time difference of arrival measurements are significantly disturbed. Therefore, a multitude of sophisticated algorithms for multipath mitigation have been investigated. The accuracy of the uplink time difference of arrival method and enhanced observed time difference method reaches values in the scale of $50 - 200\,\mathrm{m}$ [5, 9, 10, 11].

In the area of non standardized localization techniques, the *Angle of Arrival (AOA)* or *Direction Finding (DF)* method has been subject of scientific research. The main objective of the network operators is the improvement of network capacity by steering the base station beam on the active mobile phone while rejecting interference from other signal sources [12, 13]. For information transmission, the width of the beams has to be relatively broad in order to account for multipath propagation effects. Typical beam widths are in the scale of $20 - 30\,^\circ$ [12]. The accuracy of localization is reported to be in the scale of $200\,\mathrm{m}$ [14, 15].

A joint evaluation of the time difference of arrival and the angle of arrival combining the advantages of both techniques has also been investigated. Due to the increase in hardware complexity and only minor accuracy improvements, these techniques have not gained significant interest. [16, 17]

Experimental techniques based on evaluating the received signal strength of neighboring base stations have focused on the fingerprinting approach. These techniques are characterized by collecting received signal strength values at known reference points and storing these values in a database. During operation, the measured values are compared with the stored values and sophisticated algorithms for location estimation are applied. For indoor applications, a coarse classification between rooms has been demonstrated [18]. For outdoor applications, an accuracy in the scale of several hundreds of meters have been reported [19, 20]. Since these techniques rely on an extensive collection of received signal strength values before a localization is possible, interesting commercial applications have not evolved.

1.2 Localization in Search and Rescue Scenarios

The localization of mobile phones in emergency call scenarios is typically accomplished using standardized network-based localization techniques. The accuracy requirements of the E112 and E911 initiatives in the scale of $100 - 200\,\mathrm{m}$ have shown to be achievable for an adequate percentage of scenarios and propagation conditions. [21, 22, 23, 24]

Another set of scenarios, which has not been addressed by these initiatives, relates to disaster recovery and search and rescue operations. Typical scenarios comprise the localization of trapped people under collapsed structures and buildings after earthquakes and the localization of buried people after landslides or avalanches. In these scenarios, mobile phones may be used for communicating with victims and may serve as beacons for localization. The accuracy requirements are in the scale of a few meters and are very challenging from an engineering perspective. Furthermore, the case of non-functional infrastructure such as damaged base stations may have to be addressed. For this reason, this subject has regained interest in the scientific area.

First localization approaches suitable for these scenarios have recently been published. A localization procedure based on the received signal strength and the timing advance value in an emergency GSM network with satellite backhaul functionality is described in [25]. A direction finding approach for mobile phones using a portable antenna array is presented in [26]. Information on a commercially available direction finding system for GSM mobile phones can be found in [27].

In the course of research at the Institute for Electronics Engineering at the Friedrich-Alexander-University Erlangen-Nuremberg, a localization system based on the *Time Difference of Arrival (TDOA)* principle is investigated. The research is funded by the I-LOV project granted by the Federal Ministry of Education and Research (BMBF) in Germany and aims at the invention and development of advanced methods for localization and salvaging of victims in disaster scenarios. The corresponding logos are depicted in Fig. 1.3.

Figure 1.3: Logos of the I-LOV project and the German Federal Ministry of Education and Research

The main focus of this work is on the improvement of localization accuracy by employing coherent time difference of arrival estimation techniques with superior performance. This objective is accomplished by exploiting the frequency hopping capabilities of GSM signals.

1.3 Key Inventive Concept

The key inventive concept for time difference of arrival estimation is the utilization of the frequency hopping capabilities of GSM systems in order to increase the applicable bandwidth of the signals. According to the Wiener–Khinchine theorem, the accuracy of time difference of arrival estimation is directly related to the bandwidth of the applied signals [28].

For this purpose, the phase relationships between consecutive burst signals at different receiving stations have to be preserved. The presented techniques can therefore be characterized as *coherent* regarding signal acquisition and processing.

Consequently, the applicable bandwidth for time difference of arrival estimation can be increased from 200 kHz of narrowband GSM signals to the full uplink bandwidth of the corresponding GSM standard. For E-GSM 900, up to 35 MHz of bandwidth can be used for localization. The concept is illustrated in Fig. 1.4.

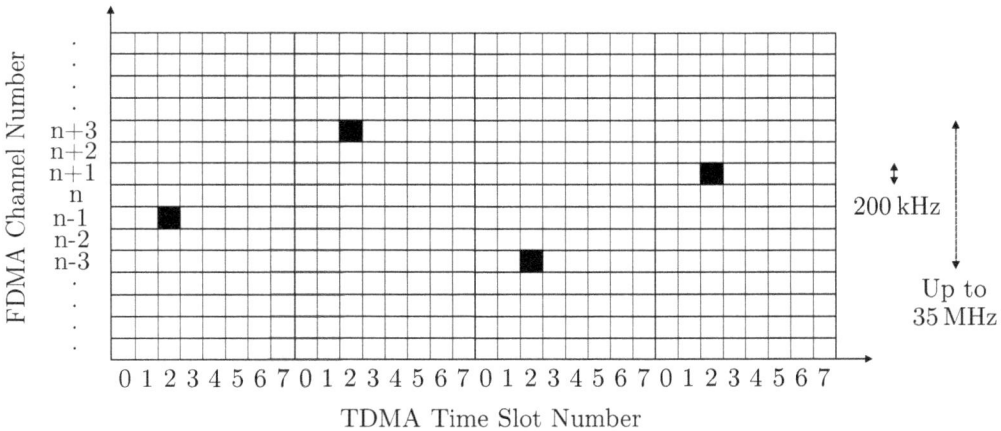

Figure 1.4: Illustration of the key inventive concept

The concept for coherent signal acquisition comprises the application of a wideband frontend covering the entire bandwidth of up to 35 MHz or the application of a narrowband frontend covering 200 kHz bandwidth of the GSM signals. In this configuration, a synchronization on the frequency hopping scheme is required. For time difference of arrival estimation, two coherent estimation techniques are proposed which make use of the coherent signal acquisition and process the acquired signals in a coherent manner.

The concepts introduced in this work provide substantial improvements compared to the state-of-the-art enabling a localization accuracy for GSM mobile phones in the scale of $5 - 10$ m. An adaption of the concepts for any frequency hopping signal source such as *Terrestrial Trunked Radio (TETRA)*, *Digital Enhanced Cordless Telecommunications (DECT)*, *IEEE 802.15.4 (ZigBee)* and *IEEE 802.15.1 (Bluetooth)* devices is possible.

1.4 Scope of this Work

The *main part* of this work is structured as follows:

In this chapter, the scientific background and a motivation for conducting research in this area have been presented. Furthermore, the key innovative concept has been described. In Chapter 2, the basic system architecture and measurement setup for time difference of arrival localization systems are introduced. The localization principles are described for a setup consisting of three receiving stations. Furthermore, an elementary measurement setup comprising two receiving stations is presented. In Chapter 3, the GSM signal characteristics are described and the modeling approach is introduced. In the scope of this work, the basic specifications of GSM are focused in order to ensure extensive applicability and compatibility of the presented techniques. The frequency hopping capabilities of GSM systems are also described in this chapter. In Chapter 4, the coherent wideband signal acquisition is presented and signal models for the acquired signals are introduced. Furthermore, the properties of the signals are investigated and pre-processing techniques are described. In Chapter 5, the coherent time difference of arrival estimation techniques are introduced. The main advantage of the algorithms is the processing of the coherent wideband signals which include the phase information of successive burst signals. Therefore, a superior estimation performance can be accomplished. In Chapter 6, the performance of the time difference of arrival estimation techniques is investigated by computer simulations. The main focus is on the noise performance and multipath separability of the techniques and a comparison with an incoherent state-of-the-art technique is conducted. In Chapter 7, the prototype system structure and measurement results are presented. The simulation results are verified and a proof-of-concept is provided. The main part of the work is concluded in Chapter 8 with a summary and outlook.

In the *appendices*, further background information related to this work is provided.

In Appendix A, the generation algorithm for frequency hopping sequences in GSM systems is described. In Appendix B, the most important logical channels in GSM and their mapping onto physical channels and resources are described. In Appendix C, the two main approaches for initiation of a frequency hopping signal transmission are presented. In Appendix D, time difference of arrival estimation techniques for narrowband GSM signals are presented. These techniques represent the state-of-the-art of time difference of arrival estimation for GSM signals today. In Appendix E, a localization system architecture for search and rescue scenarios is proposed.

2 Basic System Architecture and Measurement Setup

In this chapter, the basic system architecture and measurement setup of time difference of arrival localization systems are introduced.

The localization principles are described for a setup consisting of three receiving stations. Furthermore, an elementary measurement setup for time difference of arrival estimation comprising two receiving stations is presented.

2.1 Time Difference of Arrival Localization Systems

In the following, the structure and properties of time difference of arrival localization systems are investigated.

2.1.1 Localization Scenario

The localization scenario is characterized by a multitude of synchronized receiving stations which receive a transmitted signal from a signal source at different points in time. The signal source, e.g. a mobile phone within a GSM network, has not to be synchronized with regard to the receiving stations.

Due to these conditions, the time-of-flight of the transmitted signal from the signal source to each receiving station can not be determined directly. Since the receiving stations are synchronized, only the estimation of the time differences of arrival between pairs of receiving stations is possible.

From a mathematical point of view, these time differences of arrival represent hyperbola lines which intersect at the origin of the signal transmission. The calculation of the location of the signal source is often denoted as multilateration.

The localization of the signal source according to this procedure is also denoted as remote positioning since the location is determined by the remote receiving stations. The localization principle may also be inverted, i.e. the location may be determined by the mobile phone, which is denoted as self positioning.

In the following, the multilateration approach for a basic scenario consisting of three receiving stations based on [29] is described. The receiving stations are located at known positions (x_A, y_A), (x_B, y_B), (x_C, y_C) around a signal source of unknown position (x, y). According to the GSM terminology, the signal source is hereafter denoted as mobile station. The scenario is depicted in Fig. 2.1.

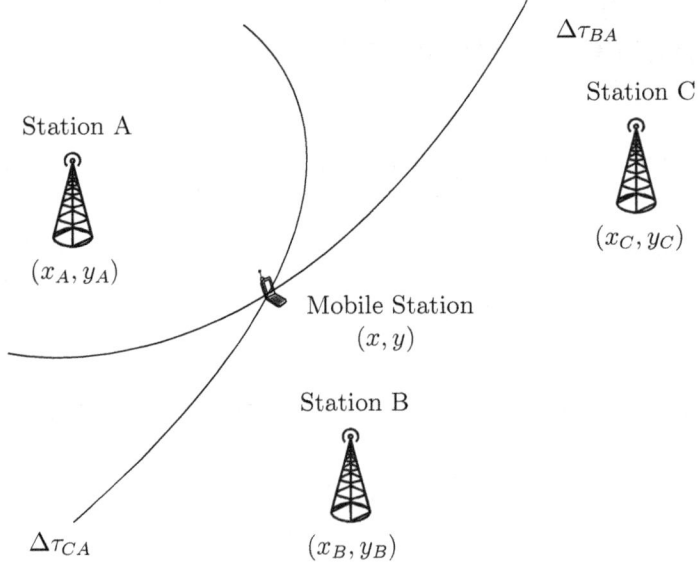

Figure 2.1: Scenario for time difference of arrival localization

With c_0 representing the speed-of-light, the times-of-flight between the signal source and the receiving stations τ_A, τ_B and τ_C can be expressed as

$$\tau_A = \frac{1}{c_0}\sqrt{(x_A - x)^2 + (y_A - y)^2} \tag{2.1}$$

$$\tau_B = \frac{1}{c_0}\sqrt{(x_B - x)^2 + (y_B - y)^2} \tag{2.2}$$

$$\tau_C = \frac{1}{c_0}\sqrt{(x_C - x)^2 + (y_C - y)^2} \tag{2.3}$$

Based on these equations, two linear independent time differences of arrival can be derived. Supposing that all time differences are calculated with respect to receiving station A and $\Delta\tau_{BA} = \tau_B - \tau_A$ and $\Delta\tau_{CA} = \tau_C - \tau_A$, the resulting equations can be expressed as

$$\Delta\tau_{BA} = \frac{1}{c_0}\left(\sqrt{(x_B - x)^2 + (y_B - y)^2} - \sqrt{(x_A - x)^2 + (y_A - y)^2}\right) \tag{2.4}$$

$$\Delta\tau_{CA} = \frac{1}{c_0}\left(\sqrt{(x_C - x)^2 + (y_C - y)^2} - \sqrt{(x_C - x)^2 + (y_C - y)^2}\right) \tag{2.5}$$

The obtained time differences of arrival can then be composed into a system vector

$$\mathbf{\Delta\tau} = [\Delta\tau_{BA}, \Delta\tau_{CA}] \tag{2.6}$$

The corresponding vector of measured time differences of arrival can be described as

$$\mathbf{\Delta\hat{\tau}} = [\Delta\hat{\tau}_{BA}, \Delta\hat{\tau}_{CA}] \tag{2.7}$$

For two-dimensional localization according to the time difference of arrival principle, a minimum of three receiving stations is required in order to obtain a minimum of two independent time difference of arrival measurements. For localization in three-dimensional space, at least four receiving stations are necessary. The presented concept can naturally be extended to an arbitrary number of receiving stations.

Depending on the measurement model, different approaches for solving the resulting equation system can be applied. For a simplified scenario without multipath propagation, the measurement model can be expressed as

$$\mathbf{\Delta\tau} = \mathbf{\Delta\hat{\tau}} + \mathbf{n} \tag{2.8}$$

with $\mathbf{n}$ denoting zero-mean *Additive White Gaussian Noise (AWGN)* with covariance matrix $\mathbf{\Sigma_n} = E\{\mathbf{n} \cdot \mathbf{n}^T\}$.

Following the weighted non-linear least squares approach, the cost function

$$\epsilon(x, y) = (\mathbf{\Delta\hat{\tau}} - \mathbf{\Delta\tau})^{\mathbf{T}} \cdot \mathbf{\Sigma_n^{-1}} \cdot (\mathbf{\Delta\hat{\tau}} - \mathbf{\Delta\tau}) \tag{2.9}$$

has to be minimized with respect to the unknown location of the signal source (x, y) yielding a location estimate of

$$(\hat{x}, \hat{y}) = \text{argmin}_{(x,y)} \ \epsilon(x, y) \tag{2.10}$$

Generally, no closed-form solutions to this non-linear and possibly over-determined optimization problem exist. Therefore, iterative approaches such as the *Gauss-Newton (GN)* algorithm, the *Steepest Descent (SD)* approach or the *Levenberg-Marquardt (LM)* algorithm may be applied [29]. An extensive overview of different multilateration algorithms is given in [30].

2.1.2 Major Challenges and Requirements

For a precise localization of the signal source, the system vector $\mathbf{\Delta\tau}$ and measurement vector $\mathbf{\Delta\hat{\tau}}$ are ideally related according to the presented AWGN measurement model. The major challenges and requirements for precise localization of the signal source can be summarized as providing measured values with minimum bias offset and minimum noise scattering.

Therefore, the synchronization of the receiving stations and the time difference of arrival estimation are of utmost importance.

The synchronization has to ensure that the timing reference is the same across all receiving stations, i.e. the clocks in all receiving stations have to be based on the same starting point in time and have to run equally fast. Therefore, the influence of jitter and timing offsets has to be minimized.

The time difference of arrival estimation is based on processing the received signals. Since the received signals have been transmitted through a radio channel, the signals are typically influenced by noise and multipath propagation. Especially the multipath components of the signal introduce a systematic bias offset which can not easily be mitigated by averaging over multiple measurements. Therefore, the identification and rejection of multipath components is of major interest. Since the resolvability of multipath components solely depends on the bandwidth of the localized signal, time difference of arrival estimation for narrowband signals is a challenging task.

The main challenges of time difference of arrival based localization systems are summarized in [31].

2.1.3 Auxiliary System Components

In the following, auxiliary system components for the operation of time difference of arrival based localization systems are presented.

Localization System for Receiving Stations

For the application of the multilateration techniques, the location of the receiving stations has to be known a-priori. For this purpose, two possibilities exist:

Global Coordinates If the receiving stations are equipped with global navigation satellite receivers, the location of the receiving stations can be determined in form of global coordinates.

Local Coordinates An alternative approach is the application of a local positioning system establishing a local coordinate system. Alternatively, the location of the receiving stations may be determined manually by the operating personnel.

The location of the receiving stations has a major influence on the attainable localization accuracy. Ideally, the localization area is characterized by a low *Geometric Dilution of Precision (GDOP)* value. [32, 33]

Synchronization Network

In order to accomplish a synchronization between the receiving stations, the stations are connected to a synchronization network as depicted in Fig. 2.2.

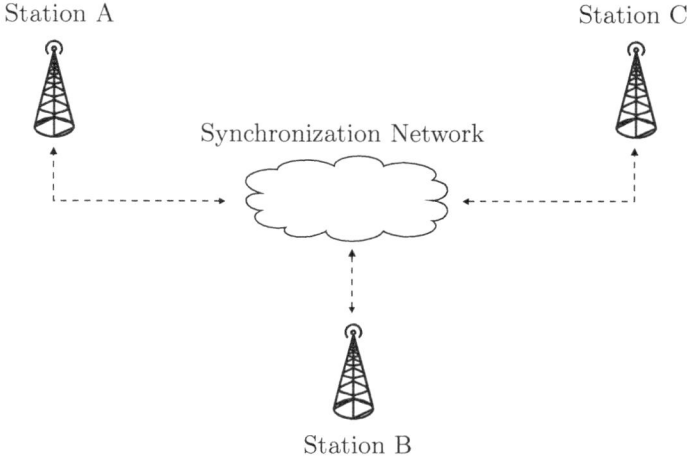

Figure 2.2: Synchronization network between the receiving stations

The synchronization refers particularly to the *Local Oscillator (LO)*, *Analog Digital Converter (ADC)* and acquisition trigger in the receiving stations.

The local oscillators in each receiving station have to be synchronized concerning the corresponding frequency and phase. The analog digital converters also have to be synchronized regarding the corresponding frequency and phase. Furthermore, the acquisition trigger instant of the signal acquisition has to be identical in all receiving stations.

The common reference signals can be generated and distributed using a wired network or a wireless network. Alternatively, the signal can be derived from a *Global Navigation Satellite System (GNSS)*.

Data Network and Evaluation Device

A data network is necessary for exchanging system parameters and received signals between the receiving stations and a central evaluation device. The evaluation device is responsible for processing the received signals, determining the time differences of arrival and control of the system components. The corresponding setup is depicted in Fig. 2.3.

Station A   Station C

Data Network

Station B   Evaluation Device

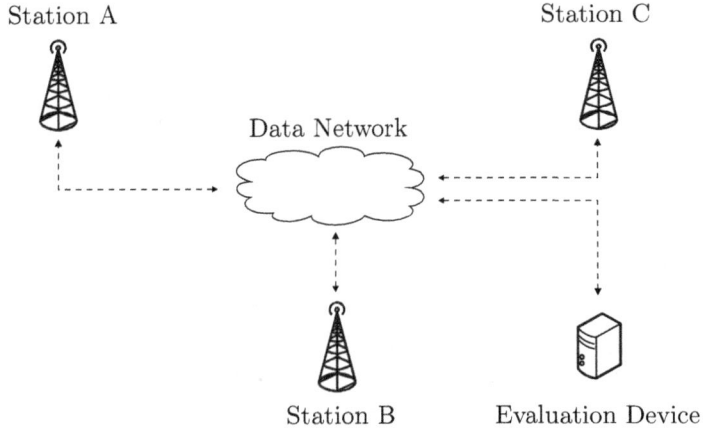

Figure 2.3: Data network and evaluation device

The data network may be realized as a wired network such as a local area network, or a wireless network such as a wireless local area network. The evaluation device may be realized as a computer or server.

2.2 Time Difference of Arrival Estimation Scenario

In this section, the time difference of arrival estimation scenario is presented. This scenario is employed for all subsequent considerations.

2.2.1 Elementary Measurement Setup

For localization purposes, at least three receiving stations for time difference of arrival measurements are required. Since the basic operation is the time difference of arrival estimation between two receiving stations, this case is focused in the following.

The actual times-of-flight τ_A and τ_B, which correspond to the line-of-sight components, relate to the corresponding distances d_A and d_B as

$$\tau_A = \frac{d_A}{c_0} \quad \text{and} \quad \tau_B = \frac{d_B}{c_0} \tag{2.11}$$

The signal is transmitted through the radio channel and is subject to noise and multipath propagation. The elementary setup for the time difference of arrival estimation is depicted in Fig. 2.4.

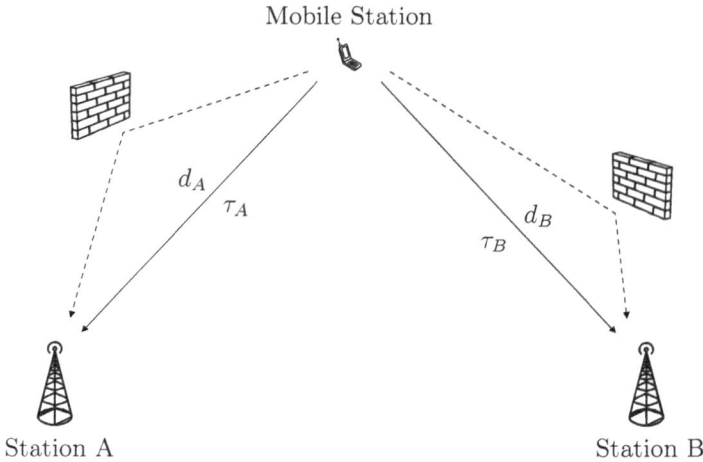

Figure 2.4: Elementary setup for the time difference of arrival estimation

The main goal of the presented techniques is the estimation of the *Time Difference of Arrival (TDOA)* between the two received signals at the corresponding receiving stations. This objective is also often denoted as *Time Delay Estimation (TDE)* in scientific literature. The time difference of arrival can be expressed as

$$\Delta\tau_{BA} = \tau_B - \tau_A \tag{2.12}$$

Since the interpretation of time differences or time delays is usually not very intuitive, these values are often expressed in terms of spatial differences or spatial delays. The corresponding value can be expressed as

$$\Delta d_{BA} = d_B - d_A = c_0(\tau_B - \tau_A) = c_0\Delta\tau_{BA} \tag{2.13}$$

For the estimation of the time difference of arrival, two different approaches are alternatively applied:

- The time delays τ_A and τ_B are estimated separately for each received signal commonly making use of a known part of the signal (e.g. a training or synchronization sequence). The time difference of arrival $\Delta\tau_{BA}$ can then be obtained by subtracting these two values.

- The time difference of arrival $\Delta\tau_{BA}$ is estimated directly using both received signals. Commonly, no part of the signals has to be known beforehand.

Many state-of-the-art techniques for time difference of arrival estimation rely on the first approach. The techniques presented in this work are entirely based on the second approach.

2.2.2 Generic Radio Channel Model

The derivation and analysis of the time difference of arrival estimation techniques requires a generic radio channel model which is introduced in the following.

The emitted signals of the mobile station propagate through the radio channel and are subject to the following effects:

Multipath Propagation The sent signals are reflected by different obstacles with different individual delays and attenuations. Therefore the received signals consist of multiple delayed and attenuated copies of the original signals.

Additive Noise The additive noise is due to physical phenomena such as thermal noise and background noise of the radio channel.

Consequently, the received signals in the two receiving stations $r_{A,RF}(t)$ and $r_{B,RF}(t)$ can be modeled as

$$r_{A,RF}(t) = s_{RF}(t) * h_{A,RF}(t) + n_{A,RF}(t) \quad \text{and} \qquad (2.14)$$

$$r_{B,RF}(t) = s_{RF}(t) * h_{B,RF}(t) + n_{B,RF}(t) \qquad (2.15)$$

with $s_{RF}(t)$ denoting the transmitted signal, $h_{A,RF}(t)$ and $h_{B,RF}(t)$ representing the corresponding radio channel impulse responses and $n_{A,RF}(t)$ and $n_{B,RF}(t)$ the noise signals in the *Radio Frequency (RF)* domain. The impulse responses are assumed to be constant over time and are considered as *Linear Time Invariant (LTI)* systems. The two noise terms are assumed to be Gaussian, white and mutually independent.

The corresponding signal flow diagram is depicted in Fig. 2.5.

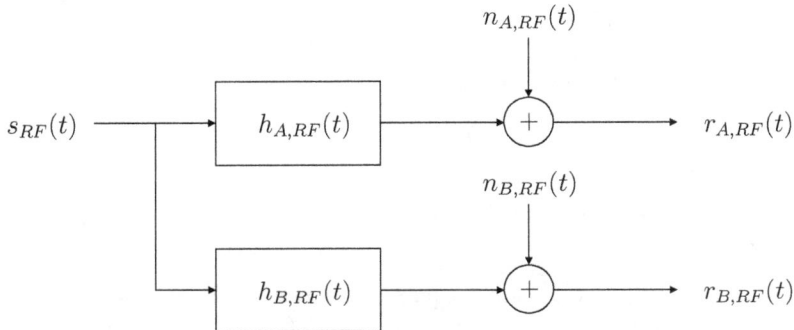

Figure 2.5: Signal flow diagram for signal transmission in RF domain

The first non-zero components of the radio channel impulse responses can be identified as the line-of-sight components, i.e.

$$\tau_A = \min \left\{ t \mid |h_{A,RF}(t)| \neq 0 \right\} \quad \text{and} \qquad (2.16)$$

$$\tau_B = \min \left\{ t \mid |h_{B,RF}(t)| \neq 0 \right\} \qquad (2.17)$$

Since the presented techniques are based on the processing of received and down-converted signals, the transmission equations are translated to the *Equivalent Complex Baseband (ECB)* domain. The received signals $r_{A,ECB}(t)$ and $r_{B,ECB}(t)$ in equivalent complex baseband domain can then be expressed as

$$r_{A,ECB}(t) = s_{ECB}(t) * h_{A,ECB}(t) + n_{A,ECB}(t) \quad \text{and} \tag{2.18}$$
$$r_{B,ECB}(t) = s_{ECB}(t) * h_{B,ECB}(t) + n_{B,ECB}(t) \tag{2.19}$$

with $s_{ECB}(t)$ denoting the transmitted signal, $h_{A,ECB}(t)$ and $h_{B,ECB}(t)$ representing the corresponding radio channel impulse responses and $n_{A,ECB}(t)$ and $n_{B,ECB}(t)$ the noise signals in equivalent complex baseband domain.

The transformation rule for the transmitted signal is given by

$$s_{ECB}(t) = \frac{1}{\sqrt{2}}(s_{RF}(t) + j\mathcal{H}\{s_{RF}(t)\})e^{-j2\pi f_{LO}t} \tag{2.20}$$

with $\mathcal{H}\{\cdot\}$ representing the Hilbert transform. The frequency f_{LO} is the corresponding local oscillator or transformation frequency. The normalization factor $1/\sqrt{2}$ guarantees a power-invariant transformation of the considered signals from the radio frequency domain to the equivalent complex baseband domain. [34]

The transformation rule for the noise terms is based on the same equation. In the case of white noise, the real valued noise in radio frequency domain with spectral density $\mathcal{N}_0/2$ is transformed to complex valued noise in equivalent complex baseband domain with spectral density $\mathcal{N}_0$. [34]

The corresponding transformation rule for the impulse responses can be expressed as

$$h_{A,ECB}(t) = \frac{1}{2}(h_{A,RF}(t) + j\mathcal{H}\{h_{A,RF}(t)\})e^{-j2\pi f_{LO}t} \quad \text{and} \tag{2.21}$$
$$h_{B,ECB}(t) = \frac{1}{2}(h_{B,RF}(t) + j\mathcal{H}\{h_{B,RF}(t)\})e^{-j2\pi f_{LO}t} \tag{2.22}$$

The normalization factor $1/2$ guarantees a power-invariant transformation of systems from the radio frequency domain to the equivalent complex baseband domain. [34]

In an ideal scenario without multipath propagation and noise, the radio channels in the RF domain can be modeled as

$$h_{A,RF}(t) = \delta(t - \tau_A) \quad \text{and} \quad h_{B,RF}(t) = \delta(t - \tau_B) \tag{2.23}$$

The corresponding radio channel impulse responses in the equivalent complex baseband domain can then be expressed as

$$h_{A,ECB}(t) = \delta(t - \tau_A)e^{-j2\pi f_{LO}\tau_A} \quad \text{and} \tag{2.24}$$
$$h_{B,ECB}(t) = \delta(t - \tau_B)e^{-j2\pi f_{LO}\tau_B} \tag{2.25}$$

under the condition of non-existing signal components for $f < -f_{LO}$ [35]. Therefore, a signal transmitted over an ideal radio channel is characterized by a delay and phase rotation of its complex envelope.

3 GSM Signal Characteristics and Modeling

In this chapter, the GSM signal characteristics are described and the modeling approach is introduced. In the scope of this work, the basic specifications of GSM are focused in order to ensure extensive applicability and compatibility of the presented techniques.

The properties of extended frequency bands such as *Digital Communication System 1800 (DCS 1800)* and *Personal Communications Service 1900 (PCS 1900)* are considered. The main concepts can also be adapted to enhanced modulation schemes and structures of evolved GSM services such as *General Packet Radio Service (GPRS)* and *Enhanced Data Rates for GSM Evolution (EDGE)*.

3.1 Physical Layer Structure

In the following, an introduction to the physical layer of GSM systems is provided. The description is based on [36] and [37].

3.1.1 Modulation

The basic modulation technique employed in GSM systems is *Gaussian Minimum Shift Keying (GMSK)* which is a a non-linear modulation format. It belongs to the family of *Continuous Phase Modulations (CPM)* with constant envelope. The carrier-modulated radio frequency signal can be expressed as

$$s_{GMSK,RF}(t) = \sqrt{\frac{2E_b}{T_b}} \cos(2\pi f_c t + \phi(t) + \phi_0) \tag{3.1}$$

where E_b is the energy per modulating bit, T_b is the modulation period and f_c is the carrier frequency. $\phi(t)$ represents the time-varying phase of the carrier and ϕ_0 its initial phase. [34]

Transforming this signal to equivalent complex baseband representation with local oscillator frequency $f_{LO} = f_c$ then yields the corresponding *Baseband (BB)* signal

$$s_{GMSK,BB}(t) = \sqrt{\frac{E_b}{T_b}} e^{j(\phi(t)+\phi_0)} \tag{3.2}$$

The time-varying phase – which carries the information – can be expressed as follows:

$$\phi(t) = \frac{\pi}{2T_b} \int_0^t \sum_{i=0}^{I-1} d_i g(\tau - iT_b) \, \mathrm{d}\tau \tag{3.3}$$

where $d_i \in \{-1, 1\}$ is the modulating data sequence of length I and $g(t)$ represents a pulse shaping function.

The modulating data sequence $d_i \in \{-1, 1\}$ which inputs the modulator is a differentially encoded version of the actual transmitted bit sequence $b_i \in \{0, 1\}$. The relationship can be expressed as

$$d_i = 1 - 2(b_i \oplus b_{i-1}) \tag{3.4}$$

where $\oplus$ denotes modulo-2 addition and $b_{-1} = 1$ by definition.

The function $g(t)$ can be interpreted as a frequency pulse which is transformed into a phase pulse by integration. For the GMSK modulation, the shape of this frequency pulse $g(t)$ is defined as a convolution of a Gaussian function $h(t)$ and a rectangular function $\mathrm{rect}(t)$ according to

$$g(t) = h(t) * \mathrm{rect}(\frac{t}{T_b}) \tag{3.5}$$

with

$$h(t) = \frac{1}{\sqrt{2\pi}\sigma T_b} \mathrm{e}^{-t^2/(2\sigma^2 T_b^2)} \quad \text{where} \quad \sigma = \frac{\sqrt{\ln 2}}{2\pi T_b B} \tag{3.6}$$

and

$$\mathrm{rect}\left(\frac{t}{T_b}\right) = \begin{cases} 1/T_b & \text{for } |t| < T_b/2 \\ 0 & \text{otherwise} \end{cases} \tag{3.7}$$

The corresponding phase pulse $q(t)$ is obtained by integration of the frequency pulse $g(t)$ given as:

$$q(t) = \int_0^t g(\tau) \, \mathrm{d}\tau \tag{3.8}$$

An important parameter which characterizes the shape of the pulse is the time-bandwidth product $T_b B$ where B denotes the 3 dB bandwidth of the Gaussian function. For GSM systems, this product is standardized as $T_b B = 0.3$. The frequency pulse $g(t)$ and phase pulse $q(t)$ are illustrated in Fig. 3.1 and Fig. 3.2.

In GSM systems, the modulation period is $T_b = 48/13 \, \mu s \approx 3.7 \, \mu s$. This corresponds to a bit rate of $1/T_b = 270.83 \, \mathrm{kBit/s}$ for data transmission without coding. Since the duration of the pulses is not limited to one modulation period T_b, the signal is characterized as a partial response signal. The non-linear modulation scheme can be decomposed into linear components as described in [38].

The modulation technique employed in GSM systems is defined in [39].

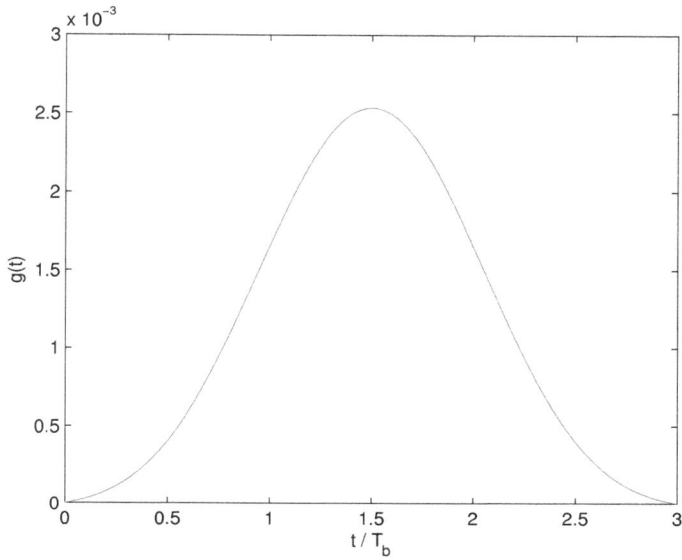

Figure 3.1: Frequency pulse $g(t)$ for $T_b B = 0.3$

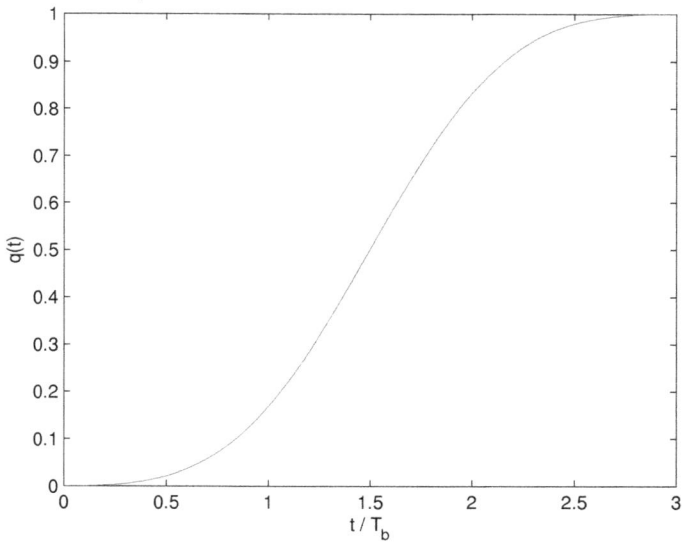

Figure 3.2: Phase pulse $q(t)$ for $T_b B = 0.3$

3.1.2 Burst Types and Structures

In GSM systems, the modulation signal is divided into burst signals of finite duration. There are several different kinds of bursts:

Normal Bursts Normal bursts are used in uplink and downlink transmission and are the most common type of bursts. They are employed for transmitting information on *Traffic Channels (TCHs)* and *Control Channels (CCHs)*.

Access Bursts Access bursts are only used in uplink transmission for gaining access to a base transceiver station. They are used for establishing a call or performing a location update. The corresponding channel is denoted as *Random Access Channel (RACH)*.

Frequency Correction Bursts Frequency correction bursts are only used in downlink transmission and enable frequency synchronization of the mobile station to the base transceiver station. The corresponding channel is denoted as *Frequency Correction Channel (FCCH)*.

Synchronization Bursts Synchronization bursts are only used in downlink transmission and enable time synchronization of the mobile station to the time frame of the base transceiver station. The corresponding channel is denoted as *Synchronization Channel (SCH)*.

Dummy Burst Dummy bursts are similar to normal bursts but contain no useful data. They are used in downlink transmission when no other bursts are to be transmitted.

Since the localization system is based on time difference of arrival estimation of uplink signals, only normal bursts and access bursts are of further interest. Therefore, these types are focused in the following.

Structure of Normal Bursts

The normal burst comprises 148 symbols with an overall time of $T_{Normalburst} = 148T_b \approx 546.5\,\mu s$. The bit structure is depicted in Fig. 3.3.

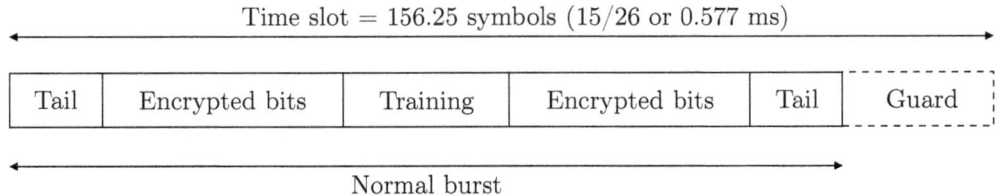

Time slot = 156.25 symbols (15/26 or 0.577 ms)

Tail	Encrypted bits	Training	Encrypted bits	Tail	Guard

Normal burst

Figure 3.3: Bit structure of normal bursts

The two blocks of encrypted bits consist of 58 symbols each. The two symbols adjacent to the Training sequence (one symbol in each block) are special signaling symbols and are denoted as stealing flags. The remaining 57 symbols in each block contain the transmitted user data.

The training sequence is centered in the middle part of the burst and comprises 26 symbols. It is used for channel estimation and equalization purposes and therefore has to be known by the receiver. There are eight different training sequences which are designated by their training sequence code. The assignment of a training sequence code to the corresponding training sequence bits is defined in [40].

The tail bits at the start and end of each burst comprise 3 symbols each and are always set to logical '0' in order to fill the time span during which the transmitter power is ramped up or down. Moreover, the zero bits are needed for demodulation purposes.

During the guard period of 8.25 symbols length, no signal is transmitted. Since GSM employs a time division multiple access scheme, this period provides protection against misalignment and ensures a collision-free burst transmission.

The definition of training sequences and tail bits can be found in [40]. The burst structure is defined in [40, 41].

Structure of Access Bursts

The access burst comprises 88 symbols with an overall time of $T_{Accessburst} = 88T_b \approx 324.9\,\mu s$. The bit structure is depicted in Fig. 3.4.

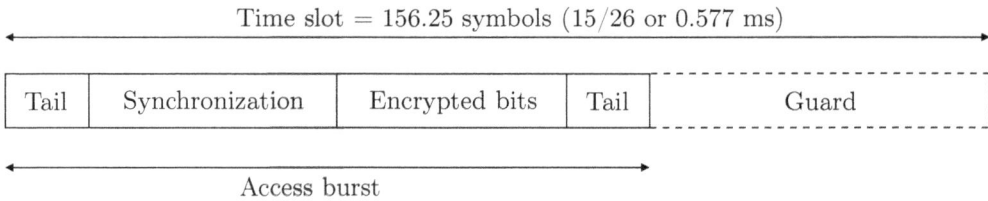

Figure 3.4: Bit structure of access bursts

The encrypted bits contain information about the mobile station which are required for the random access procedure and comprise 36 symbols.

The synchronization sequence is used for the detection of the unsynchronized mobile station and is similar to the training sequence of normal bursts. It consists of 41 symbols and is known by the base transceiver station. The increased length compared to the training sequence of normal bursts provides an additional gain in the correlator of the receiver and thus increases detection probability.

At the start and end of the burst there are tail bits of length 8 and 3 symbols, respectively. The first 8 symbols consist of a predefined bit sequence. The ending 3 symbols are all logical '0'.

The guard period of 68.25 symbols length is much longer compared to other burst types in order to reduce the probability of collision of several unsynchronized mobile stations on the random access channel and to prevent collisions with adjacent time slots.

The definition of synchronization sequences and tail bits can be found in [40]. The burst structure is defined in [40, 41].

3.1.3 Power and Spectral Characteristics

In this section, the power characteristics and spectral characteristics of GSM burst signals are summarized.

Maximum Power Level

The maximum power level of a GSM mobile station depends on the power class. The corresponding values for GSM 850, GSM 900, DCS 1800 and PCS 1900 are summarized in Tab. 3.1.

Power class	GSM 850 & GSM 900	DCS 1800	PCS 1900
1	–	1 W (30 dBm)	1 W (30 dBm)
2	8 W (39 dBm)	0.25 W (24 dBm)	0.25 W (24 dBm)
3	5 W (37 dBm)	4 W (36 dBm)	2 W (33 dBm)
4	2 W (33 dBm)	–	–
5	0.8 W (29 dBm)	–	–

Table 3.1: Power classes of GSM systems

During normal operation, these values are maximum power levels. Depending on the received signal strength at the base transceiver station, an adaptive power control may be applied. The corresponding levels are defined in [42].

For localization purposes, signal transmission at maximum levels is desirable. Therefore, the adaptive power control is supposed to be disabled and signal transmission at maximum power levels is assumed.

Power Ramping

Another aspect is the transmitted power level versus time which is also denoted as power ramping. It describes the on- and off-switching characteristics of the transmitter.

The required power mask for GMSK normal bursts and access bursts is depicted in Fig. 3.5. Some minor deviations in power levels are defined in [42].

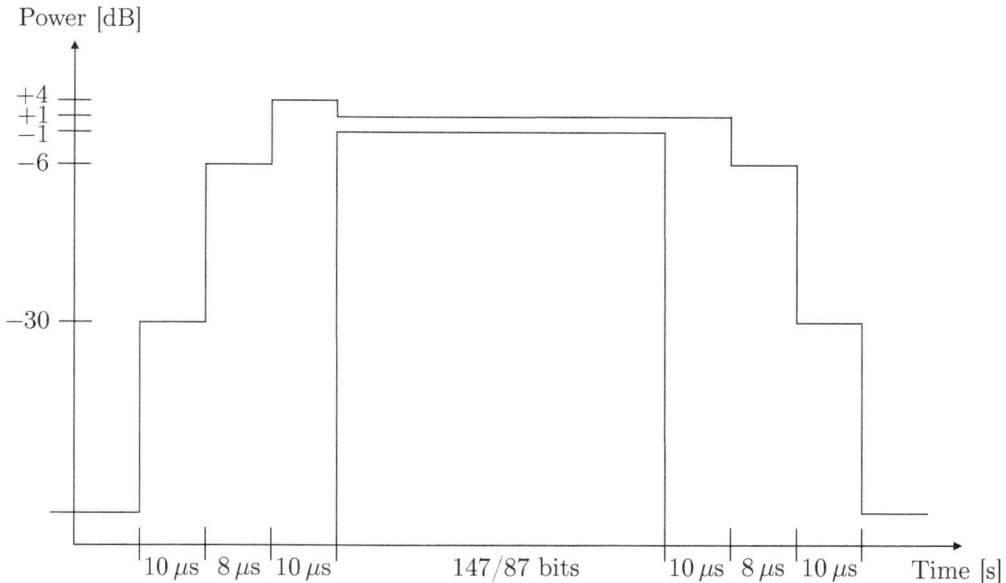

Figure 3.5: Power level versus time for GMSK normal bursts and access bursts

Within a time of 28 μs, the transmitter has to ramp up or down the power level. During the on-state the power has to be constant within a ±1 dB interval.

More detailed information about the power characteristics is provided in [42].

Mean Power Spectrum

The shape of the mean power spectrum is determined by the modulation scheme and the corresponding modulation parameters. In GSM systems, the main power of the signal is concentrated on a bandwidth of approximately 200 kHz around the carrier frequency.

An important aspect is the fact that the spectrum of a random finite duration burst signal differs from the mean spectrum. The mean power spectrum corresponds to a modulated signal of infinite duration and can be interpreted as a maximum envelope for a random burst signal.

The mean power spectrum of a single-carrier GSM signal is depicted in Fig. 3.6.

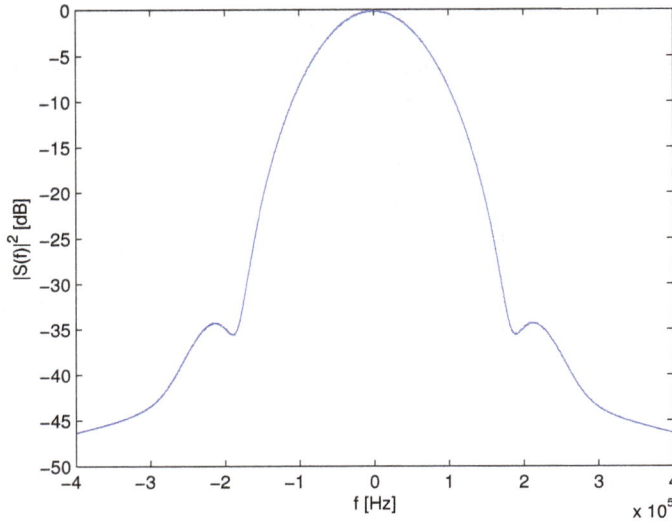

Figure 3.6: Mean power spectrum of a single-carrier GSM signal

3.1.4 Multiple Access and Duplexing

For providing multiple access on the radio channel, a multiple access scheme is required. Furthermore uplink and downlink transmissions have to be separated and therefore a duplexing technique is necessary.

For multiple access, a combination of

- *Time Division Multiple Access (TDMA)* and
- *Frequency Division Multiple Access (FDMA)*

in conjunction with a frequency hopping scheme is applied. [36]

For duplexing, a combination of

- *Time Division Duplex (TDD)* and
- *Frequency Division Duplex (FDD)*

is employed for minimizing the hardware requirements in the mobile station. [36]

Alternatively, GSM can also be considered as a mixed TDMA/FDD system using an additional *Frequency Hopped Multiple Access (FHMA)* technique. [37]

In the following, the applied multiple access and duplexing schemes are presented in more detail.

Frequency Bands and Channel Structure

In GSM systems, uplink and downlink transmissions are carried out in different frequency bands. The most common frequency bands used worldwide are shown in Tab. 3.2. Herein, the term E-GSM 900 refers to the extended GSM 900 band which includes the standard GSM 900 band.

Frequency Band	Uplink [MHz]	Downlink [MHz]	Bandwidth [MHz]
GSM 850	$824 - 849$	$869 - 894$	25
E-GSM 900	$880 - 915$	$925 - 960$	35
DCS 1800	$1710 - 1785$	$1805 - 1880$	75
PCS 1900	$1850 - 1910$	$1930 - 1990$	60

Table 3.2: Common frequency bands of GSM systems

The overall bandwidth is different for each frequency band and is of major interest with regard to the presented techniques for time difference of arrival estimation.

In available bandwidth is divided into channels of 200 kHz width. Each channel is identified by the *Absolute Radio Frequency Channel Number (ARFCN)*. The relationship between the absolute radio frequency channel number and the carrier frequency in the uplink and downlink band is given in Tab. 3.3.

Frequency Band	Uplink [MHz]	Downlink [MHz]	ARFCN
GSM 850	$824.2 + 0.2(n - 128)$	$\ldots + 45$	$128 \leq n \leq 251$
E-GSM 900	$890 + 0.2n$	$\ldots + 45$	$0 \leq n \leq 124$
	$890 + 0.2(n - 1024)$	$\ldots + 45$	$975 \leq n \leq 1023$
DCS 1800	$1710.2 + 0.2(n - 512)$	$\ldots + 95$	$512 \leq n \leq 885$
PCS 1900	$1850.2 + 0.2(n - 512)$	$\ldots + 80$	$512 \leq n \leq 810$

Table 3.3: Relationship between the ARFCN and carrier frequencies

Obviously the downlink frequencies can directly be derived from the uplink frequencies by adding a constant value. This value is the duplex frequency in terms of the frequency division duplex approach. A given absolute radio frequency channel number therefore defines a duplex frequency pair.

The available frequency bands and their properties are defined in [42].

Time Frame Structure

In time domain, one TDMA frame consists of 8 time slots containing the data bursts. Each time slot has a duration of 156.25 symbols, i.e. $T_{Timeslot} = 15/26\,\text{ms} \approx 0.577\,\text{ms}$. Therefore the TDMA frame takes $T_{Frame} = 120/26\,\text{ms} \approx 4.615\,\text{ms}$.

Every mobile phone is allocated a specific time slot within the TDMA frame, also denoted as *Time slot Number (TN)*. The structure is depicted in Fig. 3.7.

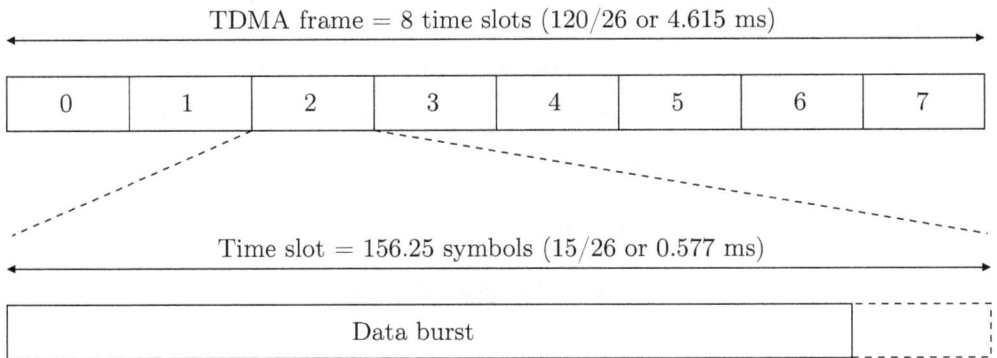

Figure 3.7: Structure of a TDMA frame

Since the arrival time of a sent data burst also depends on the distance between the mobile station and the base transceiver station, the correct timing of the signals relative to the TDMA frame has to be ensured by a special mechanism. This is accomplished by the *Timing Advance (TA)* value which describes the shift in time of signal transmission to meet the required timing conditions. Since the timing advance is automatically adapted to an adequate value, a correct alignment of the bursts relative to the TDMA frames is assumed. This procedure is also denoted as adaptive frame synchronization or adaptive frame alignment [36].

The time slot numbers for uplink and downlink transmissions are shifted in time by 3 slots. Consequently, the mobile station does not have to send and receive simultaneously and therefore does not need a special duplex unit. This procedure realizes the time division duplex approach.

For higher layers of the GSM protocol, the TDMA frames are grouped into higher frame structures. A multiframe consists of 26 TDMA frames for traffic channels or 51 TDMA frames for control channels. These multiframes are subsequently grouped into a superframe. A superframe comprises 26 51-multiframes or 51 26-multiframes, i.e. 1,326 TDMA frames. The highest frame structure is the hyperframe consisting of 2,048 superframes or 2,715,648 TDMA frames.

For synchronization of the current TDMA frame into the frame hierarchy, every individual TDMA frame is identified by a *Frame Number (FN)* which is broadcasted to the mobile stations by the base transceiver station.

More details on the time frame structure can be found in [40].

Frequency Hopping Capability

For providing diversity on the transmission links, a frequency hopping scheme is optionally applied in GSM systems.

In a frequency hopping system, the carrier frequency is changed during signal transmission. The employed scheme in GSM is a slow frequency hopping scheme, i.e. one or more symbols are transmitted in the time interval between the frequency hops.

Employing a frequency hopping scheme mitigates the following effects:

Multipath Fading Depending on the radio channel characteristics, the transmitted signal might experience frequency-selective fading, i.e. some frequency intervals show severe attenuation and no signal transmission is possible.

Interference Interfering signals introduce distortion to the transmitted signal and may negatively affect the transmission performance.

Changing the carrier frequency inherently changes the fading patterns for a given scenario, therefore introducing frequency diversity. Furthermore frequency hopping causes different signals to interfere with the transmitted signal at different times, therefore introducing interference diversity. [43]

Moreover, unauthorized signal interception is impeded since the transmitter and receiver need a-priori knowledge of the used frequency hopping scheme.

In GSM systems, the carrier frequency is constant for every individual burst and is changed in successive TDMA frames. Therefore the hopping period equals the TDMA frame duration and the corresponding hopping rate is $1/T_{Frame} \approx 216.7\,\text{Hops/s}$.

There are two basic forms of frequency hopping schemes in GSM systems:

Cyclic Hopping The hopping frequencies are gradually cycled.

Pseudo-Random Hopping The hopping frequencies are chosen according to a pseudo-random frequency hopping scheme.

The available hopping frequencies are taken from the *Cell Allocation (CA)* which is assigned to every cell within a GSM network. The cell allocation contains up to 64 different channels. The frequencies used in the hopping sequence pattern are drawn from the *Mobile Allocation (MA)* table which is a subset of the cell allocation. Since the bandwidth of a typical GSM signal exceeds the channel spacing of 200 kHz, in normal operation no adjacent channels are used for signal transmission.

The entries in the mobile allocation are numbered by the *Mobile Allocation Index (MAI)*. The entry of the mobile allocation table at which the hopping sequence begins is called the *Mobile Allocation Index Offset (MAIO)*. The applied hopping scheme is characterized by the *Hopping Sequence Number (HSN)* where $HSN = 0$ is cyclic hopping and $1 \leq HSN \leq 63$ are pseudo-random schemes. Usually, all mobile stations within a GSM cell use the same hopping sequence number and are assigned different values for the mobile allocation index offset. Thus, signal collisions between the mobile stations are not possible. The hopping sequences are also said to be orthogonal inside one cell.

The cyclic and pseudo-random hopping schemes are illustrated in Fig. 3.8. Cyclic hopping is shown as filled boxes and pseudo-random hopping as shaded boxes.

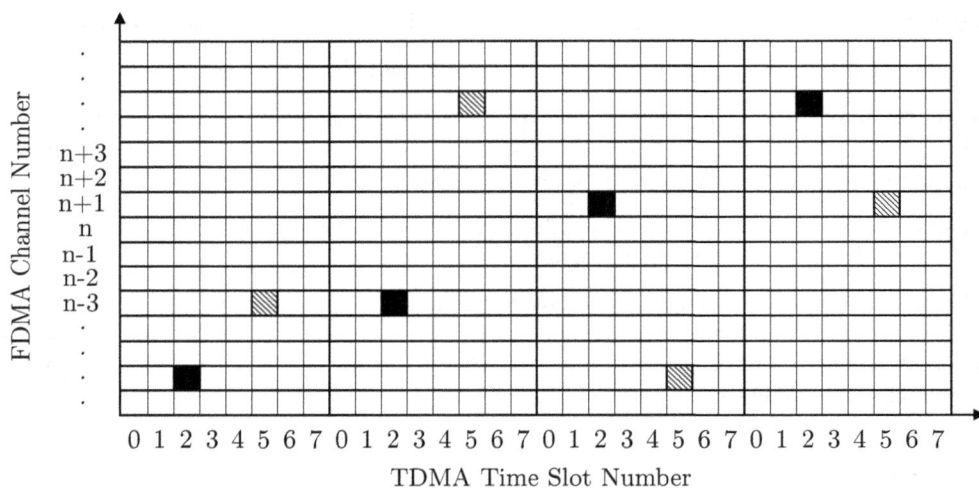

Figure 3.8: Cyclic and pseudo-random frequency hopping schemes

For the implementation of frequency hopping at the base transceiver station, there are two possibilities:

Baseband Hopping Every transceiver uses a fixed carrier frequency and therefore the mobile station hops across different transceivers.

Synthesized Hopping Every transceiver is able to change its carrier frequency and therefore the mobile station stays at the same transceiver while hopping.

This implies that for baseband hopping, the number of frequencies is limited by the number of transceivers in the base transceiver station. For synthesized hopping, these restrictions do not apply. [44]

3.2 Modeling of GSM Signals

In this section, the mathematical modeling approach for single burst signals and frequency hopping signals is described.

3.2.1 Single Burst Signal Model

A single burst signal is modeled according to the specifications of the GSM physical layer. The main focus lies on the modeling of normal bursts and access bursts, since these are the only types used in uplink signal transmission.

The model for a single burst signal in baseband representation is given as

$$s_{Burst,BB}(t, \mathbf{b_n}) = \frac{s_{GMSK,BB}(t, \mathbf{b_n})}{|s_{GMSK,BB}(t, \mathbf{b_n})|} e^{-j\phi_0} = e^{j\phi(t,\mathbf{b_n})} \quad \text{with} \quad 0 \le t \le T_{Burst} \quad (3.9)$$

The amplitude has been normalized in order to provide a unit-free signal representation. The cancellation of the initial phase ϕ_0 ensures that every considered single burst signal starts with zero phase.

The function $\phi(t, \mathbf{b_n})$ describes the time-varying phase of the complex signal for burst n and constitutes a complete representation of the burst signal with $\phi(0, \mathbf{b_n}) = 0$.

The vector $\mathbf{b_n}$ symbolizes the modulating bit sequence for burst n which has different structure and content for normal bursts and access bursts. In case of normal bursts, the tail bits, the training sequence and the stealing flags are considered as constant values. For access bursts the tail bits and the synchronization sequence are assumed to be constant as well. In both cases the encrypted bits are modeled as random with $p('0') = p('1') = 0.5$, since they contain user data and have been error-protected and channel-coded.

The duration of the burst signal is denoted as T_{Burst} and is adapted to the applied burst type, i.e.

$$T_{Burst} = \begin{cases} T_{Normalburst} = 148T_b \approx 546.5 \, \mu s & \text{for normal bursts} \\ T_{Accessburst} = 88T_b \approx 324.9 \, \mu s & \text{for access bursts} \end{cases} \quad (3.10)$$

For simplicity the power ramping characteristics of the signal have not been included in the model, because the techniques for time difference of arrival estimation only process the active part of the burst signals.

In order to obtain a compact representation of the burst signal in the radio frequency domain, complex mixing with an exponential signal is used for the model. The carrier can be characterized by the carrier frequency $f_{c,n}$ and initial phase angle $\phi_{c,n}$. The transmitted burst signal can then be modeled as

$$s_{Burst,RF}(t, \mathbf{b_n}) = s_{Burst,BB}(t, \mathbf{b_n}) e^{j(2\pi f_{c,n}t + \phi_{c,n})} \quad \text{with} \quad 0 \le t \le T_{Burst} \quad (3.11)$$

3.2.2 Frequency Hopping Signal Model

The model of the frequency hopping GSM signal is based on the model of single burst signals and is introduced in the following. For simplicity, a general model which covers all important features and is applicable for different burst types and channel mappings is employed.

The model for the frequency hopping signal in radio frequency domain is given as

$$s_{FH,RF}(t) = \sum_{n=0}^{N-1} s_{Burst,RF}(t, \mathbf{b_n}) * \delta(t - nT_{Frame}) \qquad (3.12)$$

$$= \sum_{n=0}^{N-1} s_{Burst,BB}(t, \mathbf{b_n}) e^{j(2\pi f_{c,n}t + \phi_{c,n})} * \delta(t - nT_{Frame})$$

$$\text{with} \quad 0 \leq t \leq NT_{Frame}$$

This equation represents N successive single burst signals with corresponding modulating bit sequences $\mathbf{b_n}$ for each burst signal n.

It is assumed that in consecutive bursts, the bit sequences $\mathbf{b_n}$ have different content regarding the encrypted bits and are therefore chosen randomly. All tail bits, the training or synchronization sequences and stealing flags are considered as constant values.

The carrier frequencies $f_{c,n}$ correspond to the applied frequency hopping scheme and have a major impact on the performance of the time difference of arrival estimation techniques.

The initial phase angles of the local oscillator $\phi_{c,n}$ take random values in consecutive bursts since the local oscillator in GSM mobile stations is turned off during unused time slots and starts up with random oscillation phase. Therefore, this sequence can not be influenced or used as a design parameter. For information transmission, this random nature is not of interest because the information bits are differentially encoded i.e. only the phase differences of consecutive symbols within one burst are important. For time difference of arrival estimation purposes, however, the random nature has to be taken in account.

Finally, the bursts are supposed to be aligned to the TDMA frame structure, i.e. allocated to a fixed time slot. Therefore each single burst signal is delayed by nT_{Frame} in time.

When considering a real frequency hopping signal, the structure and content inherently depend on the signal transmission initiation procedure. Furthermore, the mapping on logical channels might also have to be taken in account. For simplicity, it is assumed that no idle frames are present in the frequency hopping signal which may be accomplished by choosing an adequate zero starting point in time.

4 Coherent Wideband Signal Acquisition and Modeling

In the following sections, the coherent wideband signal acquisition is presented and signal models for the acquired signals are introduced. Furthermore, the properties of the signals are investigated and pre-processing techniques are described.

4.1 Wideband Signal Acquisition Techniques

In this section, the coherent acquisition of the wideband signals is described. Depending on the *Local Oscillator (LO)* configuration, the application of a wideband frontend or a narrowband frontend is possible.

4.1.1 Wideband Frontend Signal Acquisition

The goal of this configuration is the acquisition of the whole frequency hopping signal as transmitted over the radio channel. The approach is characterized by employing an acquisition bandwidth of the frontend of at least two single burst bandwidths. Ideally, the bandwidth covers the whole uplink band of the applied GSM standard. The acquisition interval is supposed to cover the whole length of the signal over several TDMA frames including empty time slots.

The local oscillator for down-conversion of the received signal is operated at a fixed frequency f_{LO} and initial phase angle ϕ_{LO}. The acquired signal can then be modeled as multiplication with the corresponding complex exponential function.

A single burst signal in equivalent complex baseband notation can be expressed as

$$
\begin{aligned}
s_{Burst,ECB}(t, \mathbf{b_n}) &= s_{Burst,RF}(t, \mathbf{b_n}) e^{-j(2\pi f_{LO}t + \phi_{LO})} \\
&= s_{Burst,BB}(t, \mathbf{b_n}) e^{j(2\pi(f_{c,n} - f_{LO})t + (\phi_{c,n} - \phi_{LO}))} \\
&= s_{Burst,BB}(t, \mathbf{b_n}) e^{j(2\pi\Delta f_n t + \Delta\phi_n)} \\
&\text{with} \quad 0 \le t \le T_{Burst}
\end{aligned}
\tag{4.1}
$$

with Δf_n denoting the frequency difference and $\Delta\phi_n$ the phase difference between the carrier and the local oscillator components.

The complete acquired frequency hopping signal can then be expressed as

$$s_{FH,ECB}(t) = \sum_{n=0}^{N-1} s_{Burst,ECB}(t, \mathbf{b_n}) * \delta(t - nT_{Frame}) \tag{4.2}$$

$$= \sum_{n=0}^{N-1} s_{Burst,BB}(t, \mathbf{b_n}) e^{j(2\pi\Delta f_n t + \Delta\phi_n)} * \delta(t - nT_{Frame})$$

$$\text{with} \quad 0 \leq t \leq NT_{Frame}$$

The frequency of the local oscillator f_{LO} may be chosen arbitrarily as long as all burst signals lie within the acquisition bandwidth of the frontend. In a preferred configuration, this frequency is chosen as mean value of all occurring carrier frequencies, i.e.

$$f_{LO} = \frac{1}{N} \sum_{n=0}^{N-1} f_{c,n} \tag{4.3}$$

yielding an acquired signal with zero mean value. Otherwise the mean value should be removed before applying the techniques for time difference of arrival estimation.

The coherent signal acquisition using a wideband frontend is depicted in Fig. 4.1.

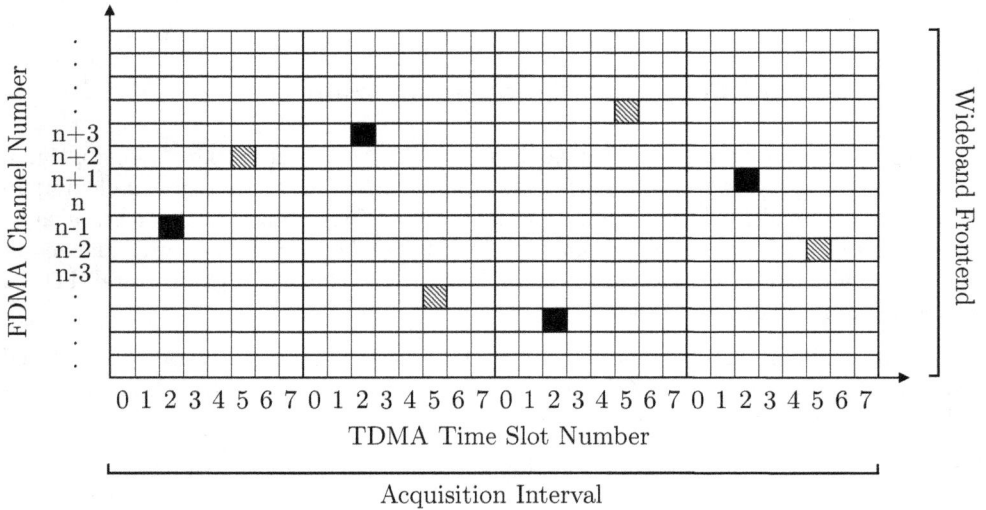

Figure 4.1: Wideband signal acquisition using a wideband frontend

The presented signal acquisition technique has several advantages:

- The frequency hopping signal is acquired without prior knowledge of the frequency hopping scheme and occurring carrier frequencies. Thus, an independent operation of the receiving stations with regard to the mobile phone network is possible.

- The hardware configuration is simple using a fixed local oscillator frequency.

- All existing signal components in time and frequency are acquired and enable a simultaneous processing and localization of multiple signal sources.

The drawbacks can be summarized as follows:

- The signal acquisition using a wideband frontend requires a sampling rate considerably higher than for single burst signals. Furthermore, the acquisition interval contains empty time slots which provide no information for the time difference of arrival estimation techniques. The data volume and required processing power are therefore high.

- The signal to noise ratio of the signal is low since undesired noise components in time and frequency are present in the acquired signal.

- In the case of multiple signal sources, the dynamic range of the frontend may be limited since the reference power level has to be adapted to the strongest signal component.

4.1.2 Narrowband Frontend Signal Acquisition

The goal of this configuration is the acquisition of individual burst signals of the frequency hopping signal. The approach is characterized by employing an acquisition bandwidth of the frontend of a single burst bandwidths independent of the overall hopping bandwidth. The acquisition intervals only comprise the active parts of the signal.

The signal acquisition using a narrowband frontend requires a coarse synchronization in time and frequency on the frequency hopping scheme, i.e. the local oscillator frequency and the acquisition intervals have to be adapted on time. This synchronization may not be confused with the synchronization requirements of local oscillators, analog digital converters and trigger signals in different receiving stations.

In this configuration, the frequency and initial phase angle of the local oscillator are dependent on the burst index n. Furthermore, it is assumed, that the local oscillator frequency equals the carrier frequency of the burst signal, i.e. $f_{LO,n} = f_{c,n}$.

A single burst signal in equivalent complex baseband notation can be expressed as

$$s_{Burst,ECB}(t, \mathbf{b_n}) = s_{Burst,RF}(t, \mathbf{b_n}) e^{-j(2\pi f_{LO,n} t + \phi_{LO,n})} \tag{4.4}$$
$$= s_{Burst,BB}(t, \mathbf{b_n}) e^{j(2\pi(f_{c,n} - f_{LO,n})t + (\phi_{c,n} - \phi_{LO,n}))}$$
$$= s_{Burst,BB}(t, \mathbf{b_n}) e^{j\Delta\phi_n}$$
$$\text{with} \quad 0 \leq t \leq T_{Burst}$$

with $\Delta\phi_n$ denoting the phase difference between the carrier and the local oscillator components.

The acquired frequency hopping signal can then be represented as a set of functions

$$\{s_{Burst,ECB}(t, \mathbf{b_0}), s_{Burst,ECB}(t, \mathbf{b_1}), \ldots, s_{Burst,ECB}(t, \mathbf{b_{N-1}})\} \qquad (4.5)$$

The coherent signal acquisition using a narrowband frontend is depicted in Fig. 4.2.

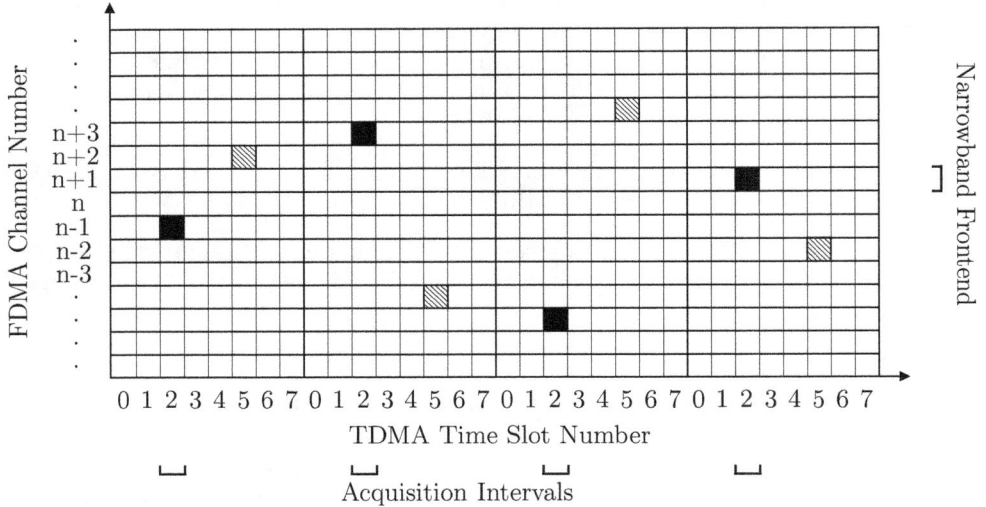

Figure 4.2: Wideband signal acquisition using a narrowband frontend

The presented signal acquisition technique has several advantages:

- The sampling rate is considerably lower compared to the wideband acquisition since only the single burst bandwidth has to be covered. Furthermore, the isolated acquisition intervals only comprise the active parts of the signal. This results in a substantially lower data volume and required processing power compared to wideband signal acquisition.

- The signal to noise ratio of the acquired signal is high since only active signal components are present.

- The dynamic range of the frontend is considerably higher compared to wideband signal acquisition.

The drawbacks can be summarized as follows:

- The signal acquisition using a narrowband frontend requires a coarse synchronization in time and frequency on the frequency hopping scheme. Therefore, the hardware complexity is increased compared to wideband signal acquisition.

- Only one signal source can be acquired and therefore localized at a time.

4.1.3 Mixed Acquisition Schemes

The presented acquisition configurations using a wideband frontend or a narrowband frontend represent complementary possibilities for the acquisition of wideband signals.

The wideband frontend signal acquisition is characterized by capturing the whole signal in time and frequency. The narrowband frontend signal acquisition is characterized by synchronization on the frequency hopping sequence in time and frequency.

For the coherent acquisition of the wideband signals, mixed acquisition schemes, as depicted in Fig. 4.3, may also be employed.

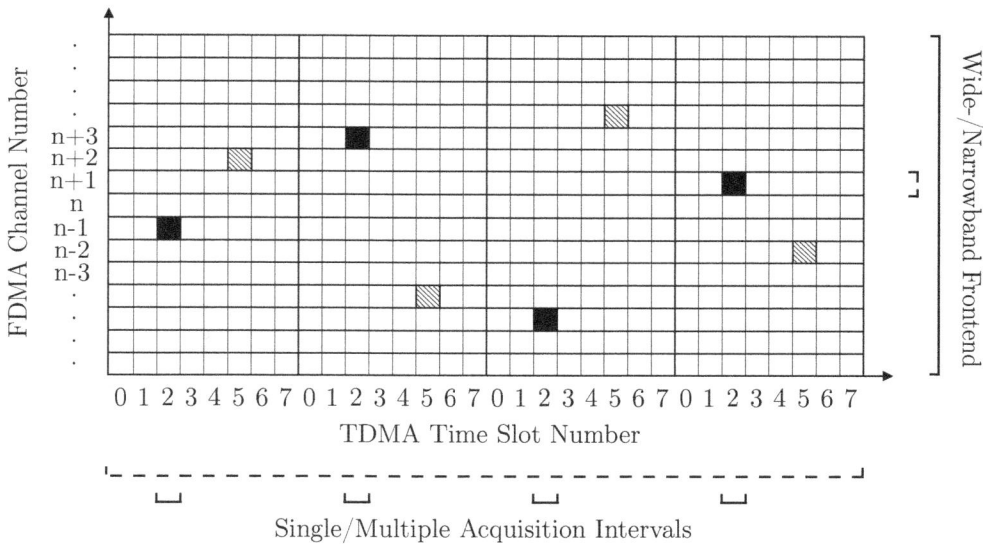

Figure 4.3: Wideband signal acquisition using mixed acquisition schemes

The first mixed acquisition scheme employs a wideband frontend with acquisition intervals focused on the active parts of the frequency hopping signal. The second mixed acquisition scheme uses a narrowband frontend acquiring the whole frequency hopping signal over time.

Both schemes require a coarse synchronization in time and frequency on the frequency hopping scheme, i.e. the local oscillator frequency and the acquisition intervals have to be adapted according to the frequency hopping scheme. As long as the synchronization requirements of local oscillators, analog digital converters and trigger signals in different receiving stations are maintained during the active parts of the signal, the mixed acquisition schemes may be applied analogously.

Since the mixed acquisition schemes offer no obvious advantage compared to the presented wideband and narrowband frontend acquisition techniques, they are not further considered in the following sections.

4.1.4 Coherence and Phase Relationships

For the coherent techniques for time difference of arrival estimation, the coherence and phase relationships of the acquired signals have to be preserved. The coherent wideband signal acquisition techniques are capable of providing these requirements assuming synchronized receiving stations.

The coherence and phase relationships between the acquired signals in receiving station A and B for consecutive burst signals are illustrated in Fig. 4.4.

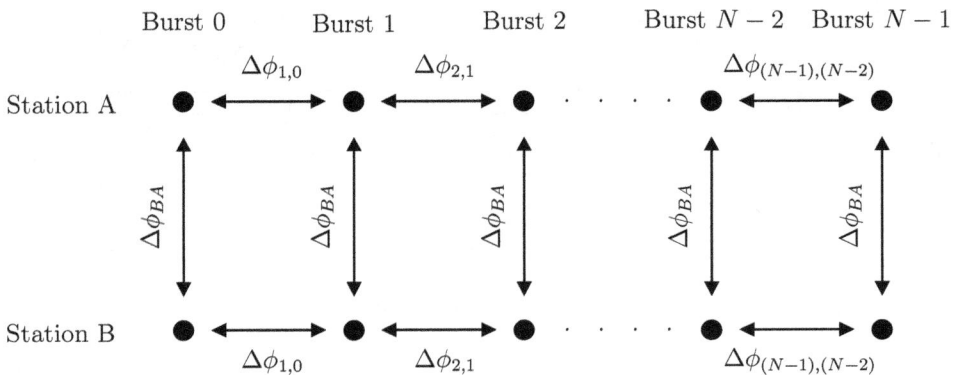

Figure 4.4: Illustrative representation of coherence and phase relationships

The value $\Delta\phi_{BA}$ characterizes the phase offset between receiving station A and B and is supposed to be constant. This assumption is assured by the synchronization modules in the corresponding receiving stations and is valid of all burst signals.

The values $\Delta\phi_{i,j}$ represent the phase offsets of adjacent burst signals i and j and are assumed to be identical for the corresponding received burst signals in receiving station A and B.

The term coherence refers to the phase preserving nature of the wideband signal acquisition and time difference of arrival estimation techniques. The narrowband frequency hopping signal can then be interpreted and evaluated as a wideband signal providing superior estimation performance.

4.2 Properties of Acquired Wideband Signals

The acquired wideband signals can be characterized by different measures which are introduced in the following sections.

4.2.1 Frequency Hopping Bandwidth

The signal hopping bandwidth characterizes the overall bandwidth of the acquired frequency hopping signal and has a major impact on the resolution of the time difference of arrival estimation techniques.

The hopping bandwidth depends on the carrier frequency sequence $f_{c,n}$ and can be calculated as

$$B_{Hopping} = \max(f_{c,n}) - \min(f_{c,n}) + 200\,\text{kHz} \tag{4.6}$$

In the case of wideband signal acquisition with $f_{c,n} = f_{LO} + \Delta f_n$, the hopping bandwidth can alternatively be calculated as

$$\begin{aligned} B_{Hopping} &= \max(f_{LO} + \Delta f_n) - \min(f_{LO} + \Delta f_n) + 200\,\text{kHz} \\ &= \max(\Delta f_n) - \min(\Delta f_n) + 200\,\text{kHz} \end{aligned} \tag{4.7}$$

The additive term of 200 kHz considers the signal energy at the edges of the covered frequency band.

4.2.2 Effective Signal Bandwidth

The effective signal bandwidth is based on the second moment of the spectrum of the acquired frequency hopping signal and has a major impact on the noise performance of the time difference of arrival estimation techniques.

The effective signal bandwidth is defined as

$$B_{Eff} = \sqrt{\frac{(2\pi)^2 \int_{-\infty}^{+\infty} f^2 |S(f)|^2 \,\mathrm{d}f}{\int_{-\infty}^{+\infty} |S(f)|^2 \,\mathrm{d}f}} \tag{4.8}$$

with $|S|^2$ representing the baseband spectral energy density of the signal. [45]

For the mean baseband spectrum of a GSM signal as shown in Fig. 3.6, the effective bandwidth can be calculated numerically yielding a value of $B_{Eff} \approx 318.1\,\text{kHz}$. For frequency hopping signals, the baseband representation is intuitively accomplished using a wideband frontend with the local oscillator frequency centered on the overall spectrum.

In order to provide a closed form approximation for the effective bandwidth of frequency hopping signals using wideband signal acquisition, it is now assumed that the energies of the individual burst signals are equal and that they are concentrated on the respective carrier frequencies. With $|\hat{S}|^2$ denoting the burst energies, the overall spectral energy density may then be expressed as

$$|S(f)|^2 = \sum_{n=0}^{N-1} |\hat{S}|^2 \delta(f - \Delta f_n) \tag{4.9}$$

The effective bandwidth can then be approximated as

$$
\begin{aligned}
B_{Eff} &= \sqrt{\frac{(2\pi)^2 \int_{-\infty}^{+\infty} f^2 \sum_{n=0}^{N-1} |\hat{S}|^2 \delta(f - \Delta f_n)\, \mathrm{d}f}{\int_{-\infty}^{+\infty} \sum_{n=0}^{N-1} |\hat{S}|^2 \delta(f - \Delta f_n)\, \mathrm{d}f}} \\[2mm]
&= \sqrt{\frac{(2\pi)^2 |\hat{S}|^2 \int_{-\infty}^{+\infty} f^2 \sum_{n=0}^{N-1} \delta(f - \Delta f_n)\, \mathrm{d}f}{|\hat{S}|^2 \int_{-\infty}^{+\infty} \sum_{n=0}^{N-1} \delta(f - \Delta f_n)\, \mathrm{d}f}} \\[2mm]
&= \sqrt{\frac{(2\pi)^2 \int_{-\infty}^{+\infty} f^2 \sum_{n=0}^{N-1} \delta(f - \Delta f_n)\, \mathrm{d}f}{\int_{-\infty}^{+\infty} \sum_{n=0}^{N-1} \delta(f - \Delta f_n)\, \mathrm{d}f}} \\[2mm]
&= \sqrt{\frac{(2\pi)^2 \sum_{n=0}^{N-1} \Delta f_n^2}{N}}
\end{aligned}
\tag{4.10}
$$

using the sampling property of the weighted delta functions. The effective bandwidth of frequency hopping signals can then be approximated using only the frequency sequence Δf_n.

4.2.3 Signal to Noise Ratio

In this section, the relationship between the signal to noise ratio using a narrowband frontend and a wideband frontend is investigated.

The signal to noise ratio which is attained by a narrowband frontend covering essentially the narrowband bandwidth can be expressed as

$$SNR_{Narrowband} = \frac{E_{Burst}/T_{Burst}}{\mathcal{N}_0 B_{Narrowband}} \tag{4.11}$$

In this equation, E_{Burst} represents the energy of the single burst signal and T_{Burst} denotes the burst duration. $\mathcal{N}_0$ is the spectral noise power density which is considered as constant over the considered frequency range and $B_{Narrowband}$ is the narrowband bandwidth.

The narrowband bandwidth is defined as the common bandwidth of GSM signals as

$$B_{Narrowband} = 200\,\text{kHz} \tag{4.12}$$

The signal to noise ratio of the acquired signal using a wideband frontend depends on the acquisition bandwidth of the system which is denoted as $B_{Wideband}$. The signal to noise ratio can then be calculated as

$$
\begin{aligned}
SNR_{Wideband} &= \frac{E_{Total}/T_{Total}}{\mathcal{N}_0 B_{Wideband}} = \frac{(NE_{Burst})/(NT_{Frame})}{\mathcal{N}_0 B_{Narrowband}(B_{Wideband}/B_{Narrowband})} \\
&= \frac{(E_{Burst}/T_{Burst})(T_{Burst}/T_{Frame})}{\mathcal{N}_0 B_{Narrowband}(B_{Wideband}/B_{Narrowband})} \\
&= SNR_{Narrowband} \frac{T_{Burst}}{T_{Frame}} \frac{B_{Narrowband}}{B_{Wideband}}
\end{aligned}
\tag{4.13}
$$

The first signal to noise ratio correction factor depends on the employed burst type. Generally, the application of normal bursts provides an increase in the signal to noise ratio of 2.26 dB compared to access bursts due to their increased length and signal energy. The numeric values for the correction factor T_{Burst}/T_{Frame} are given in Tab. 4.1.

Burst type	Linear	Logarithmic
Normal burst	0.1184	−9.3 dB
Access burst	0.0704	−11.5 dB

Table 4.1: SNR correction factor for different burst types

The second signal to noise ratio correction factor is dependent on the acquisition bandwidth of the wideband frontend. In the following table, it is assumed that the full allocated uplink bandwidth of the corresponding GSM standard is used. The numeric values for the correction factor $B_{Narrowband}/B_{Wideband}$ are summarized in Tab. 4.2.

GSM standard	Linear	Logarithmic
GSM 850	0.0080	−21.0 dB
E-GSM 900	0.0057	−22.4 dB
DCS 1800	0.0027	−25.7 dB
PCS 1900	0.0033	−24.8 dB

Table 4.2: SNR correction factor for different GSM standards

The signal to noise ratio of the acquired wideband signal is severely reduced due to the empty time slots and the large acquisition bandwidth. Using the presented burst isolation technique in conjunction with the wideband signal reconstruction technique, the signal to noise ratio using a wideband frontend can be improved significantly and reaches at best the narrowband frontend case.

4.3 Pre-Processing and Multiple Source Separation

In this section, the isolation of burst signals and the reconstruction of wideband signals are described. These pre-processing techniques enable a conversion between the acquired signals of wideband and narrowband frontends. Furthermore, a separation technique for multiple signal sources for simultaneous localization is provided.

The pre-processing of the received signals has to be performed identically in all receiving stations in order to ensure the applicability of the estimation techniques. If necessary, the relevant parameters have to be exchanged between the receiving stations beforehand.

4.3.1 Burst Isolation Technique

The acquired signal using a wideband frontend comprises a structure in time and frequency domain. The active parts of the signal are the most interesting components and are isolated in time and frequency as a first step.

A spectrogram of an acquired wideband signal is depicted in Fig. 4.5.

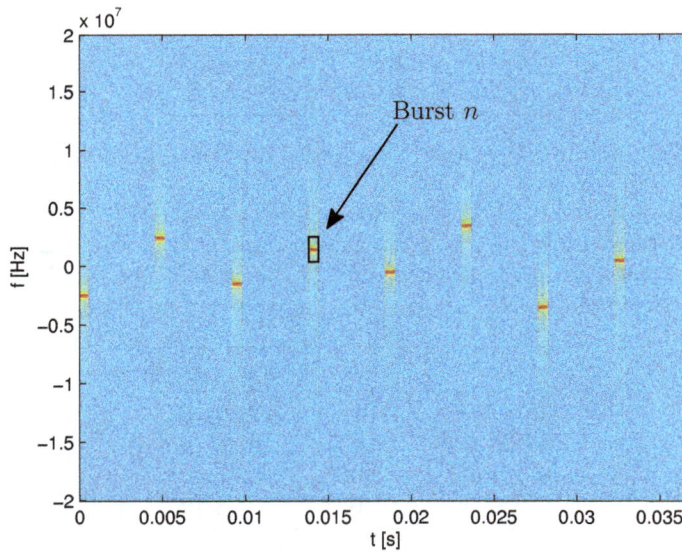

Figure 4.5: Acquired wideband signal in time and frequency

The procedures are described using the sampled version of the acquired signal

$$s_{FH,ECB}[k] = s_{FH,ECB}(kT_s) = s_{FH,ECB}(k\frac{1}{f_s}) \tag{4.14}$$

The sampling frequency is denoted as f_s and the sampling interval as T_s.

The isolation in time is based on the absolute value of the acquired signal and can be accomplished as follows:

1. Calculation of the absolute value of the signal $|s_{FH,ECB}[k]|$

2. Finding the first index k_0 which exceeds a given threshold level ν

3. Calculation of the first index of burst n by $k_n = k_0 + \lfloor nT_{Frame}f_s \rfloor$

4. Extracting the signal samples with indices $k_n \leq k \leq k_n + \lfloor T_{Burst}f_s \rfloor$

$\lfloor \cdot \rfloor$ represents the floor function and is necessary in the sampled domain. The threshold level ν could easily be determined relative to the maximum level by a multiplicative relationship e.g. $\nu = 0.8 \max(|s_{FH,ECB}[k]|)$. Of course, the outermost samples could be omitted in order to ensure that no empty samples and transient effects are included. Note that the sampling frequency is not changed by the isolation in time.

The isolation in frequency comprises a bandpass filtering process on the time isolated signal which could be implement by different means. In our case, a discrete Fourier transform based approach is employed which could be realized as follows:

1. Calculation of the spectrum using a discrete Fourier transform operation

2. Determination of the center frequency of the acquired burst $\tilde{l}$

3. Calculation of the sample window length L which corresponds to a frequency window bandwidth B_{Window}

4. Keeping the samples $\tilde{l} - L/2 \leq l \leq \tilde{l} + L/2$ unaltered and setting all remaining samples to zero

5. Calculation of the corresponding sample sequence by an inverse discrete Fourier transform operation

The center frequency of the acquired burst depends on the frequency of the local oscillator and the carrier frequency of the corresponding burst. By knowledge of the frequency hopping sequence, this value is available a-priori. Otherwise a frequency analysis could yield this value. The frequency window bandwidth B_{Window} can be chosen arbitrarily, ideally as multiples of the channel spacing of $200\,$kHz.

The discretization, especially the $\lfloor \cdot \rfloor$ operation, is necessary in the sampled domain and has no effect on the estimation algorithms as long as the received signals are pre-processed all in the same way with identical parameters (such as starting samples, window length, etc.). If necessary, these parameters have to be exchanged between the receiving stations beforehand.

The resulting isolated signal in time and frequency containing burst n is denoted

$$s_{Burst,Isolated,n}[k] = s_{Burst,Isolated,n}(kT_s) = s_{Burst,Isolated,n}(k\frac{1}{f_s}) \qquad (4.15)$$

In our implementation, the isolation of the burst signal does not affect the sampling frequency. Therefore, the isolated signal is generally highly oversampled. Furthermore, the isolated burst signal has only been bandpass filtered and therefore still resides at its original frequency band. Naturally, the technique can be refined (including zero-padding, etc.) yielding equivalent isolated burst signals at different sampling frequencies.

A spectrogram of an isolated burst signal using $B_{Window} = 400\,\text{kHz}$ is shown in Fig. 4.6.

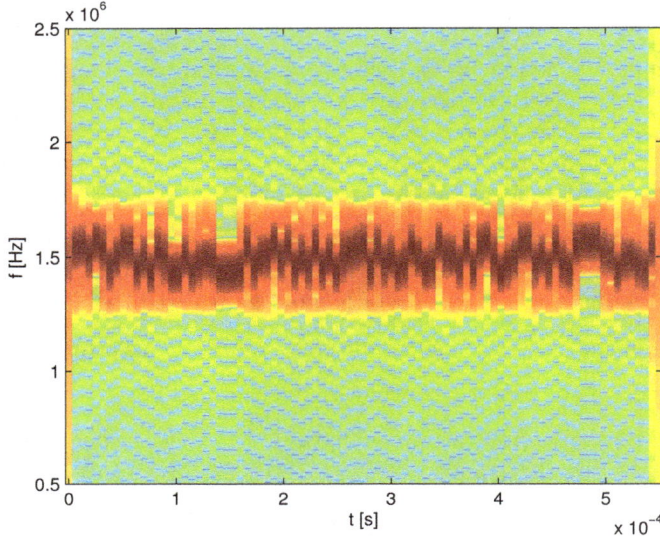

Figure 4.6: Isolated burst n in time and frequency

The transformation to narrowband representation comprises two steps:

1. Frequency shift to zero frequency

2. Decimation/Down-sampling by an adequate factor

The frequency shift can be accomplish by multiplication with a complex exponential signal yielding

$$s_{Burst,Shifted,n}[k] = s_{Burst,Isolated,n}[k]\mathrm{e}^{-j2\pi\Delta f_n kT_s} \tag{4.16}$$

The decimation can easily be accomplished by a down sampling operation since the necessary filtering has already been carried out during the isolation of the burst. The resulting signal can therefore be described as

$$s_{Burst,ECB,n}[\tilde{k}] = s_{Burst,Shifted,n}[D\tilde{k}] \tag{4.17}$$

The corresponding decimation factor is denoted as D.

4.3.2 Wideband Signal Reconstruction Technique

The wideband signal reconstruction technique is complementary to the burst isolation technique. The goal of this procedure is the generation of an equivalent wideband signal based on the acquisition of narrowband signals using a narrowband frontend.

The procedures are described using the sampled versions of the acquired signals

$$s_{Burst,ECB,n}[\tilde{k}] = s_{Burst,ECB}(\tilde{k}T_s, \mathbf{b_n}) = s_{Burst,ECB}(\tilde{k}\frac{1}{f_s}, \mathbf{b_n}) \qquad (4.18)$$

f_s represents the sampling frequency and T_s the corresponding sampling interval.

The transformation to wideband representation comprises two steps:

1. Interpolation/Up-sampling by a suitable factor

2. Frequency shift to offset frequency

As a first step, the sampling rate of the acquired signal has to be increased to the required value according to the Nyquist theorem in order to cover the bandwidth of the resulting wideband signal. For this purpose, the signal has to be interpolated which may be realized using a common up-sampling and low-pass filtering operation or a discrete Fourier transform based approach.

The interpolated signal is denoted as $s_{Burst,Interpolated,n}[k]$. Using an interpolation factor of I, the following relationship holds:

$$s_{Burst,Interpolated,n}[\tilde{k}I] = s_{Burst,ECB,n}[\tilde{k}] \qquad (4.19)$$

The interpolated signal has now to be shifted to the offset frequency where the burst is intended to reside. The corresponding frequency Δf_n can be derived from the frequency hopping scheme. The frequency shift can be accomplish by multiplication with a complex exponential signal yielding

$$s_{Burst,Shifted,n}[k] = s_{Burst,Interpolated,n}[k]e^{j2\pi\Delta f_n kT_s} \qquad (4.20)$$

The complete wideband signal can now be reconstructed by the insertion of empty time slots. The resulting signal, which is equivalent to an acquisition using a wideband frontend, can be expressed as

$$s_{FH,ECB}[k] = [s_{Burst,Shifted,0}[\cdot], \underbrace{0,0,0,\ldots}_{\lfloor 7T_{Timeslot}f_s \rfloor}, s_{Burst,Shifted,N-1}[\cdot], \underbrace{0,0,0,\ldots}_{\lfloor 7T_{Timeslot}f_s \rfloor}] \qquad (4.21)$$

The generated signal contains no additional noise components and may be used for time difference of arrival estimation.

4.3.3 Multiple Source Separation Technique

The presented pre-processing techniques provide an interesting means for the separation of multiple signal sources. These techniques are very useful for signal acquisition using wideband frontends since they enable a simultaneous localization of multiple sources during a single acquisition process.

The separation is accomplished by isolating the burst signals of the corresponding source and reconstructing a specific wideband signal. The time difference of arrival estimation techniques can then be applied independently on the corresponding signal sources. As an example, the separation of two signal sources is depicted in Fig. 4.7.

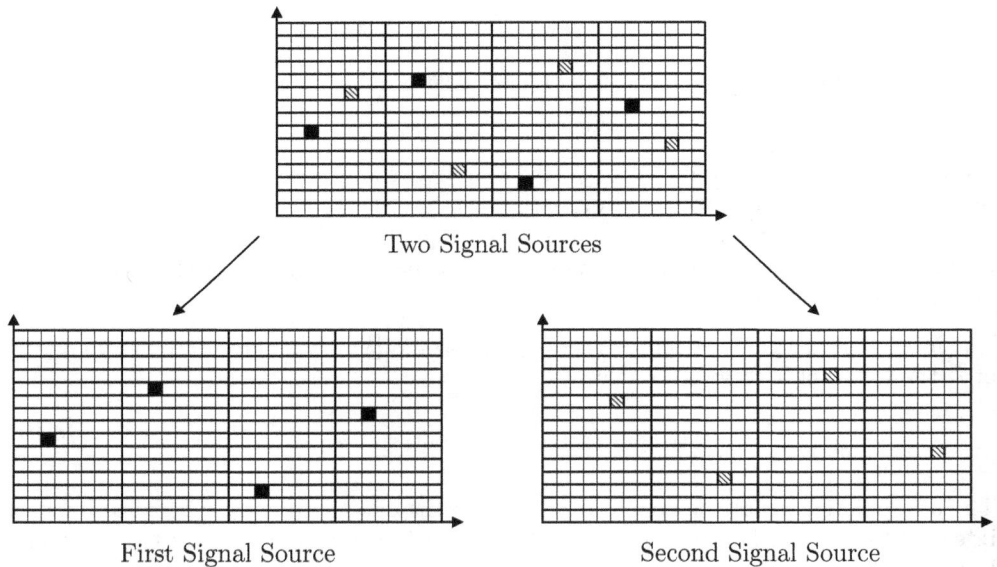

Figure 4.7: Separation technique for two signal sources

For a unique assignment of the burst signals to a specific mobile station, the employed time slot and frequency hopping sequence have to be know a-priori. The frequency information may not be necessary however, if all signal sources use different time slots.

In the case of a narrowband frontend, the signal acquisition has to be synchronized on the time slot and frequency hopping sequence. The isolation of a single signal source is therefore inherently accomplished. A separation of different signal sources is not possible in this case.

5 Coherent Time Difference of Arrival Estimation Techniques

In the following sections, the coherent time difference of arrival estimation techniques are introduced. The main advantage of the algorithms is the processing of coherent wideband signals which include the phase information of successive burst signals.

Similar approaches for different applications have been presented for IEEE 802.15.4 (ZigBee) devices [46, 47, 48] and stepped frequency radar systems [49, 50, 51, 52, 53].

5.1 Wideband Crosscorrelation Technique

The first coherent technique for time difference of arrival estimation is based on crosscorrelation and is described in the following sections.

5.1.1 Definition of the Crosscorrelation Function

Crosscorrelation is a common approach for determining the time delay between two signals. Commonly, the crosscorrelation operation is performed on single narrowband burst signals which results in broad correlation peaks and significant errors in time difference of arrival estimation.

The presented approach is based on performing the crosscorrelation directly on the acquired wideband signals which cover a wider portion of the frequency spectrum. Therefore this approach is denoted as wideband crosscorrelation.

The *Crosscorrelation Function (CCF)* between the two received signals $r_{A,ECB}(t)$ and $r_{B,ECB}(t)$ can be expressed as follows:

$$\text{CCF}(\Delta\tau) = r_{A,ECB}(t) \star r_{B,ECB}(t)\Big|_{\Delta\tau} = \int_{-\infty}^{\infty} r_{A,ECB}^*(t) r_{B,ECB}(t + \Delta\tau)\, \mathrm{d}t \qquad (5.1)$$

where $\Delta\tau$ denotes the time delay between the two signals and $\star$ represents the crosscorrelation operator.

The time delay that maximizes the absolute value of the crosscorrelation function $\Delta\hat{\tau}_{BA}$ is interpreted as estimation for the actual time difference of arrival, i.e.

$$\Delta\hat{\tau}_{BA} = \mathrm{argmax}_{\Delta\tau} |\mathrm{CCF}(\Delta\tau)| \tag{5.2}$$

Of course, the crosscorrelation function can also be defined in terms of a spatial delay $\Delta d = c_0 \Delta\tau$ instead of a time delay $\Delta\tau$. Both definitions are equivalent and are used in the following sections.

5.1.2 Decomposition of the Crosscorrelation Function

In the following, the crosscorrelation function $\mathrm{CCF}(\Delta\tau)$ is decomposed into a signal component and a radio channel dependent component. The transmitted signal $s_{ECB}(t)$ may represent the complete frequency hopping signal $s_{FH,ECB}(t)$ or a single burst signal $s_{Burst,ECB}(t)$.

$$
\begin{aligned}
\mathrm{CCF}(\Delta\tau) &= r_{A,ECB}(t) \star r_{B,ECB}(t) \Big|_{\Delta\tau} = r_{A,ECB}^*(-t) * r_{B,ECB}(t) \Big|_{\Delta\tau} \tag{5.3} \\
&= \Big[\big(s_{ECB}^*(-t) * h_{A,ECB}^*(-t) + n_{A,ECB}^*(-t) \big) * \\
&\qquad \big(s_{ECB}(t) * h_{B,ECB}(t) + n_{B,ECB}(t) \big) \Big] \Big|_{\Delta\tau} \\
&= \Big[s_{ECB}^*(-t) * h_{A,ECB}^*(-t) * s_{ECB}(t) * h_{B,ECB}(t) \\
&\qquad + s_{ECB}^*(-t) * h_{A,ECB}^*(-t) * n_{B,ECB}(t) \\
&\qquad + s_{ECB}(t) * h_{B,ECB}(t) * n_{A,ECB}^*(-t) \\
&\qquad + n_{A,ECB}^*(-t) * n_{B,ECB}(t) \Big] \Big|_{\Delta\tau} \\
&= \Big[s_{ECB}^*(-t) * s_{ECB}(t) * h_{A,ECB}^*(-t) * h_{B,ECB}(t) \Big] \Big|_{\Delta\tau} \\
&= \Big[\big(s_{ECB}(t) \star s_{ECB}(t) \big) * \big(h_{A,ECB}(t) \star h_{B,ECB}(t) \big) \Big] \Big|_{\Delta\tau}
\end{aligned}
$$

In this expression, the relationship between crosscorrelation and convolution has been used and it has been assumed that convolutions involving noise terms are zero.

Obviously, the shape of the crosscorrelation function is determined by the autocorrelation of the original signal $s_{ECB}(t)$ and the crosscorrelation between the two radio channel impulse responses $h_{A,ECB}(t)$ and $h_{B,ECB}(t)$.

In an ideal scenario, the radio channel impulse responses just contain a single delta function at the corresponding delay, i.e.

$$h_{A,ECB}(t) = \delta(t - \tau_A)\mathrm{e}^{-j2\pi f_{LO}\tau_A} \quad \text{and} \tag{5.4}$$

$$h_{B,ECB}(t) = \delta(t - \tau_B)\mathrm{e}^{-j2\pi f_{LO}\tau_B} \tag{5.5}$$

In this case, the crosscorrelation function of the two impulse responses just contains a single delta function with a phase shift at the time difference between the two respective delays.

$$
\begin{aligned}
h_{A,ECB}(t) \star h_{B,ECB}(t) &= \left(\delta(t - \tau_A) e^{-j2\pi f_{LO}\tau_A} \right) \star \left(\delta(t - \tau_B) e^{-j2\pi f_{LO}\tau_B} \right) \\
&= \left(\delta^*(-t - \tau_A) e^{j2\pi f_{LO}\tau_A} \right) * \left(\delta(t - \tau_B) e^{-j2\pi f_{LO}\tau_B} \right) \\
&= \left(\delta^*(-t - \tau_A) * \delta(t - \tau_B) \right) e^{j2\pi f_{LO}\tau_A} e^{-j2\pi f_{LO}\tau_B} \\
&= \left(\delta(t - \tau_A) \star \delta(t - \tau_B) \right) e^{-j2\pi f_{LO}(\tau_B - \tau_A)} \\
&= \delta(t - \Delta\tau_{BA}) e^{-j2\pi f_{LO}\Delta\tau_{BA}}
\end{aligned}
\tag{5.6}
$$

Consequently, the autocorrelation function of $s_{ECB}(t)$ is shifted by $\Delta\tau_{BA}$ and exhibits a phase rotation by $-2\pi f_{LO}(\tau_B - \tau_A)$.

Since the autocorrelation function of any technically interesting signal exhibits its maximum value at zero delay, the time difference of arrival between the two received signals can be estimated by maximum search on the crosscorrelation function $CCF(\Delta\tau)$.

In a more complex scenario, the crosscorrelation function of the two impulse responses may contain additional delta functions close to the interesting value $\Delta\tau_{BA}$. In order to resolve the different delay values, a sharp autocorrelation peak of $s_{ECB}(t)$ is desirable.

5.1.3 Ambiguity Function of Frequency Hopping Signals

The shape of the autocorrelation function of $s_{ECB}(t)$ has a major influence on the crosscorrelation function $CCF(\Delta\tau)$ and is investigated in the following.

In radar technology, the autocorrelation function of a radar signal (including shifts in time and frequency) is denoted as ambiguity function [54]. In the scope of this work, only shifts in time are considered.

In the following, it is assumed that a frequency hopping signal is used as originating signal, i.e.

$$
s_{ECB}(t) \stackrel{\text{Def.}}{=} s_{FH,ECB}(t)
\tag{5.7}
$$

The ambiguity function $s_{FH,ECB}(t) \star s_{FH,ECB}(t)$ can then be derived as follows:

$$
\left(\sum_{n=0}^{N-1} s_{Burst,BB}(t, \mathbf{b_n}) e^{j(2\pi\Delta f_n t + \Delta\phi_n)} * \delta(t - nT_{Frame}) \right) \star
\tag{5.8}
$$

$$
\left(\sum_{m=0}^{M-1} s_{Burst,BB}(t, \mathbf{b_m}) e^{j(2\pi\Delta f_m t + \Delta\phi_m)} * \delta(t - mT_{Frame}) \right)
$$

Since only small delays between the two signals are considered, only corresponding burst signals with the same index $n = m$ contribute to the ambiguity function. Therefore the expression can be simplified to

$$\sum_{n=0}^{N-1} \left(s_{Burst,BB}(t, \mathbf{b_n}) e^{j(2\pi\Delta f_n t + \Delta\phi_n)} \right) \star \left(s_{Burst,BB}(t, \mathbf{b_n}) e^{j(2\pi\Delta f_n t + \Delta\phi_n)} \right) \tag{5.9}$$

The term inside the summation represents the autocorrelation function of individual frequency-shifted burst signals. This function can be reduced by using the integral representation of the correlation function

$$\int_{-\infty}^{\infty} s_{Burst,BB}^{*}(t, \mathbf{b_n}) e^{-j(2\pi\Delta f_n t + \Delta\phi_n)} s_{Burst,BB}(t + \Delta\tau, \mathbf{b_n}) e^{j(2\pi\Delta f_n (t+\Delta\tau) + \Delta\phi_n)} \, dt \tag{5.10}$$

which can be simplified to

$$\int_{-\infty}^{\infty} s_{Burst,BB}^{*}(t, \mathbf{b_n}) s_{Burst,BB}(t + \Delta\tau, \mathbf{b_n}) e^{j2\pi\Delta f_n \Delta\tau} \, dt \tag{5.11}$$

$$= e^{j2\pi\Delta f_n \Delta\tau} \int_{-\infty}^{\infty} s_{Burst,BB}^{*}(t, \mathbf{b_n}) s_{Burst,BB}(t + \Delta\tau, \mathbf{b_n}) \, dt$$

$$= e^{j2\pi\Delta f_n \Delta\tau} \left(s_{Burst,BB}(t, \mathbf{b_n}) \star s_{Burst,BB}(t, \mathbf{b_n}) \right)$$

Finally, the ambiguity function $s_{FH,ECB}(t) \star s_{FH,ECB}(t)$ can be stated as follows:

$$\sum_{n=0}^{N-1} e^{j2\pi\Delta f_n \Delta\tau} \left(s_{Burst,BB}(t, \mathbf{b_n}) \star s_{Burst,BB}(t, \mathbf{b_n}) \right) \tag{5.12}$$

5.1.4 Shape of the Ambiguity Function

The shape of the ambiguity function depends basically on the autocorrelation functions of the burst signals and the length and structure of the frequency sequence Δf_n. In the following, the influence of these two components is investigated.

Autocorrelation Functions of Single Narrowband Burst Signals

At first, the properties of the autocorrelation functions of the narrowband burst signals

$$s_{Burst,BB}(t, \mathbf{b_n}) \star s_{Burst,BB}(t, \mathbf{b_n}) \tag{5.13}$$

are analyzed. Since these functions are complex-valued, the absolute value and phase value have to be considered. In Fig. 5.1 and Fig. 5.2 these two values are depicted for ten randomly chosen normal bursts.

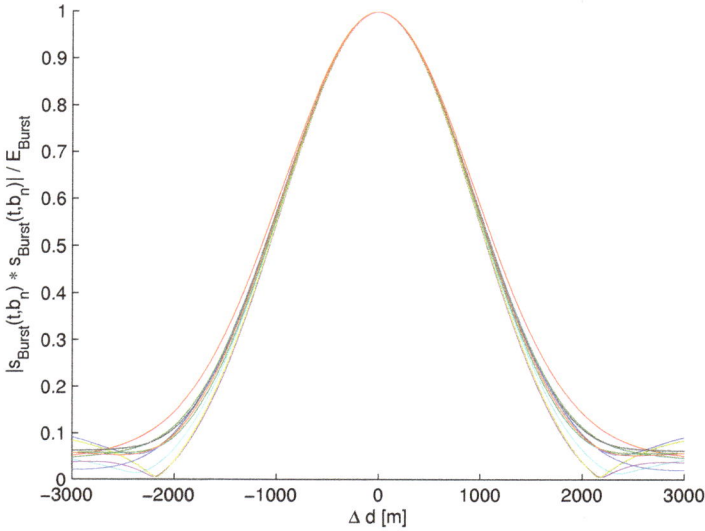

Figure 5.1: Absolute value of the autocorrelation function of ten normal burst signals

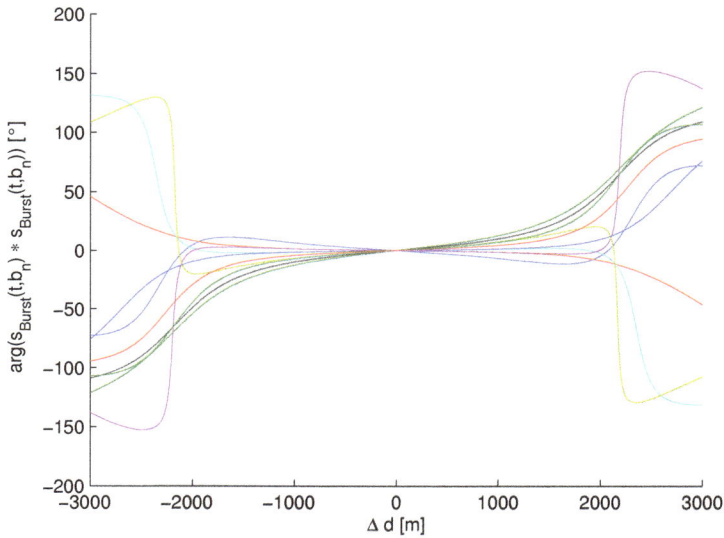

Figure 5.2: Phase value of the autocorrelation function of ten normal burst signals

As can be seen, the autocorrelation functions are not identical which is due to the different bit sequences. In the case of small delay values, e.g. $-500\,\text{m} \le \Delta d \le 500\,\text{m}$, the differences between the functions can be assumed to be negligible.

The width of the autocorrelation peak (full width half maximum) is about $2.3\,\text{km}$. Since only radio channel components exceeding this distance can be separated from the desired time difference of arrival value Δd_{BA}, a crosscorrelation technique based on single narrowband burst signals is not suitable for accurate time difference of arrival estimation.

The given results are also valid for the autocorrelation functions of access bursts. The fact that access bursts are shorter in time, does not have any influence on the shape of the autocorrelation function. The width of the correlation peak is only dependent on the occupied bandwidth which is the same for normal and access bursts.

Multi-Carrier Response Function for Arbitrary Frequency Sequences

In order to simplify the mathematical derivations, it is now assumed that the autocorrelation functions of all burst signals are identical. This approximation is reasonable for small delay values but also yields representative results for larger delays.

The ambiguity function $s_{FH,ECB}(t) \star s_{FH,ECB}(t)$ can then be expressed as

$$\sum_{n=0}^{N-1} e^{j2\pi \Delta f_n \Delta \tau} \left(s_{Burst,BB}(t) \star s_{Burst,BB}(t) \right) \tag{5.14}$$

$$= \left(s_{Burst,BB}(t) \star s_{Burst,BB}(t) \right) \sum_{n=0}^{N-1} e^{j2\pi \Delta f_n \Delta \tau}$$

The influence of the frequency sequence Δf_n on the ambiguity function can now be described by the function

$$\sum_{n=0}^{N-1} e^{j2\pi \Delta f_n \Delta \tau} \tag{5.15}$$

which can be interpreted as time domain response of a multi-carrier signal using N subcarriers and corresponding frequencies Δf_n. This term is complex-valued and is denoted as multi-carrier response function in the following.

As an example, the multi-carrier response function for the following frequency sequence is investigated:

$$\Delta f_n = \{400\,\text{kHz}, -1.6\,\text{MHz}, 800\,\text{kHz}, -2\,\text{MHz}, 1.2\,\text{MHz}, -400\,\text{kHz}, 2\,\text{MHz}, -1.2\,\text{MHz}\}$$

The absolute value and the complex representation have been chosen in order to illustrate the correspondence between the dominant components and the corresponding phase values and are depicted in Fig. 5.3 and Fig. 5.4.

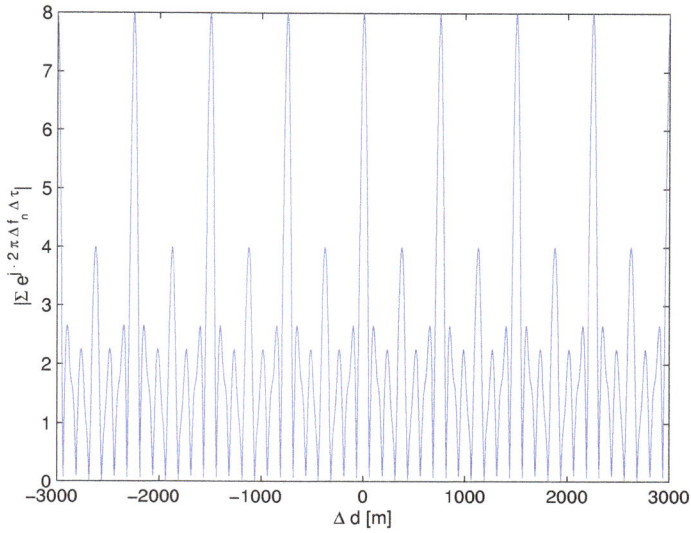

Figure 5.3: Absolute value of an arbitrary multi-carrier response function

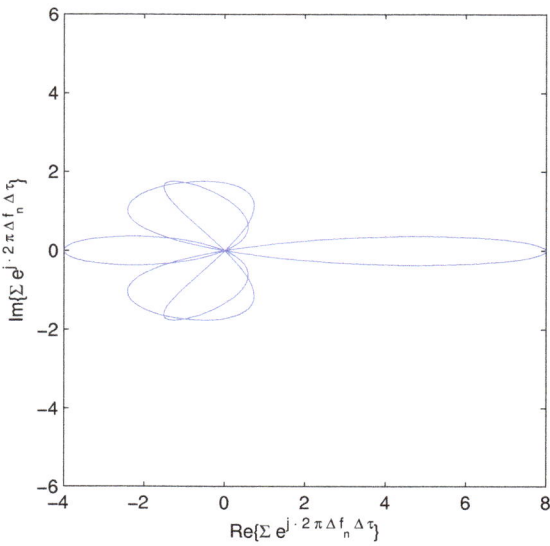

Figure 5.4: Complex representation of an arbitrary multi-carrier response function

This function shows narrow peaks and exhibits a periodic structure. The strongest peaks occur when the subcarriers interfere constructively. In this example, the absolute values of the dominant peaks equal N and the phase values are zero. All other peaks exhibit non-zero phase values.

From a mathematical point of view, the order of the frequency entries in the frequency sequence Δf_n is not relevant. Only the overall occupancy of the spectrum is of interest.

In this example, the gaps in the spectrum are $400\,\text{kHz}$ or $800\,\text{kHz}$ which corresponds to $c_0/400\,\text{kHz} = 750\,\text{m}$ and $c_0/800\,\text{kHz} = 375\,\text{m}$. The most dominant components are located at multiples of these distances.

Multi-Carrier Response Function for Linear Frequency Sequences

In order to gain further insight into the properties of this function, it is now assumed that the entries in the frequency sequence increase or decrease linearly. This approach enables further mathematical simplifications and provides closed-form expressions.

The linear frequency sequence can be expressed in two equivalent forms:

$$\Delta f_n = n\Delta f_{Step} - \frac{(N-1)\Delta f_{Step}}{2} \quad \text{with} \quad 0 \le n \le N-1 \tag{5.16}$$

$$\Delta f_k = k\Delta f_{Step} \quad \text{with} \quad -\frac{N-1}{2} \le k \le \frac{N-1}{2} \tag{5.17}$$

Using these relationships, the multi-carrier response function can be simplified as follows:

$$\sum_{n=0}^{N-1} e^{j2\pi\Delta f_n\Delta\tau} = \sum_{k=-\frac{N-1}{2}}^{\frac{N-1}{2}} e^{j2\pi\Delta f_k\Delta\tau} = \sum_{k=-\frac{N-1}{2}}^{\frac{N-1}{2}} e^{j2\pi k\Delta f_{Step}\Delta\tau} \tag{5.18}$$

For even N this term can be expressed as

$$\sum_{k=-\frac{N-1}{2}}^{\frac{N-1}{2}} e^{j2\pi k\Delta f_{Step}\Delta\tau} = \sum_{l=\frac{1}{2}}^{\frac{N-1}{2}} \left(e^{j2\pi l\Delta f_{Step}\Delta\tau} + e^{-j2\pi l\Delta f_{Step}\Delta\tau} \right) \tag{5.19}$$

$$= \sum_{l=\frac{1}{2}}^{\frac{N-1}{2}} 2\cos(2\pi l\Delta f_{Step}\Delta\tau)$$

For odd N this term can be simplified to

$$\sum_{k=-\frac{N-1}{2}}^{\frac{N-1}{2}} e^{j2\pi k\Delta f_{Step}\Delta\tau} = 1 + \sum_{l=1}^{\frac{N-1}{2}} \left(e^{j2\pi l\Delta f_{Step}\Delta\tau} + e^{-j2\pi l\Delta f_{Step}\Delta\tau} \right) \qquad (5.20)$$

$$= 1 + \sum_{l=1}^{\frac{N-1}{2}} 2\cos(2\pi l\Delta f_{Step}\Delta\tau)$$

The multi-carrier response function for linear frequency sequences can be described as a harmonic superposition of cosine functions. Interestingly, the function and is real-valued for all N. Furthermore the function contains a direct component for odd N.

As an example, the multi-carrier response function for $N = 8$ and $\Delta f_{Step} = 400\,\mathrm{kHz}$ is depicted in Fig. 5.5.

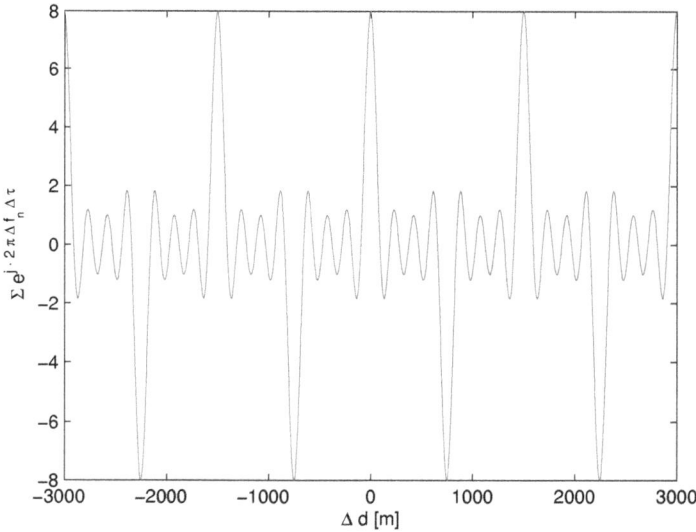

Figure 5.5: Illustration of a real-valued multi-carrier response function

The delay values with constructive superposition are characterized by

$$\Delta\tau = m\frac{1}{\Delta f_{Step}} \quad \text{or} \quad \Delta d = m\frac{c_0}{\Delta f_{Step}} \quad \text{with} \quad m \in \mathbb{Z} \qquad (5.21)$$

For even N, the corresponding values at these delays can be calculated as

$$\sum_{l=\frac{1}{2}}^{\frac{N-1}{2}} 2\cos(2\pi l \Delta f_{Step}\Delta\tau) = \sum_{l=\frac{1}{2}}^{\frac{N-1}{2}} 2\cos(2\pi lm) = \begin{cases} N & \text{for even } m \\ -N & \text{for odd } m \end{cases} \tag{5.22}$$

For odd N, the corresponding values at these delays are

$$1 + \sum_{l=1}^{\frac{N-1}{2}} 2\cos(2\pi l \Delta f_{Step}\Delta\tau) = 1 + \sum_{l=1}^{\frac{N-1}{2}} 2\cos(2\pi lm) = N \tag{5.23}$$

These mathematical derivations have been based on the assumption of a linearly increasing or decreasing frequency sequence. As already noted, the order of the entries in the frequency sequence does not have any influence on the shape since all signal components are additively superimposed. The obtained results are therefore also valid for non-linear frequency sequences which have been generated by permutations of linear frequency sequences.

Overall Ambiguity Function

In the case of a simplified scenario comprising identical autocorrelation functions of all burst signals, the overall ambiguity function can be calculated as product of the narrowband burst autocorrelation function and the multi-carrier response function according

$$s_{FH,ECB}(t) \star s_{FH,ECB}(t) = \left(s_{Burst,BB}(t) \star s_{Burst,BB}(t)\right) \sum_{n=0}^{N-1} e^{j2\pi\Delta f_n\Delta\tau} \tag{5.24}$$

In the case of a real scenario, the separation of the burst autocorrelation functions and the multi-carrier response function is not possible. The ambiguity function is rather given by the coherent summation according

$$s_{FH,ECB}(t) \star s_{FH,ECB}(t) = \sum_{n=0}^{N-1} e^{j2\pi\Delta f_n\Delta\tau} \left(s_{Burst,BB}(t,\mathbf{b_n}) \star s_{Burst,BB}(t,\mathbf{b_n})\right) \tag{5.25}$$

For illustration of the overall shapes, the absolute values of the simplified and real ambiguity function for a linear frequency sequence using $N = 8$ and $\Delta f_{Step} = 400\,\text{kHz}$ are compared in Fig. 5.6 and Fig. 5.7. For simplicity, the absolute values have been normalized to the total energy of the signal NE_{Burst}.

The shape of the ambiguity function is mainly determined by the dominant components of the multi-carrier response function. The autocorrelation function of the narrowband burst signals can be interpreted as envelope of the ambiguity function which is sampled at certain delay values. Interestingly, the main shape of the ambiguity function remains preserved using identical or non-identical burst autocorrelation functions. Some minor deviations are caused by the coherent summation in real scenarios.

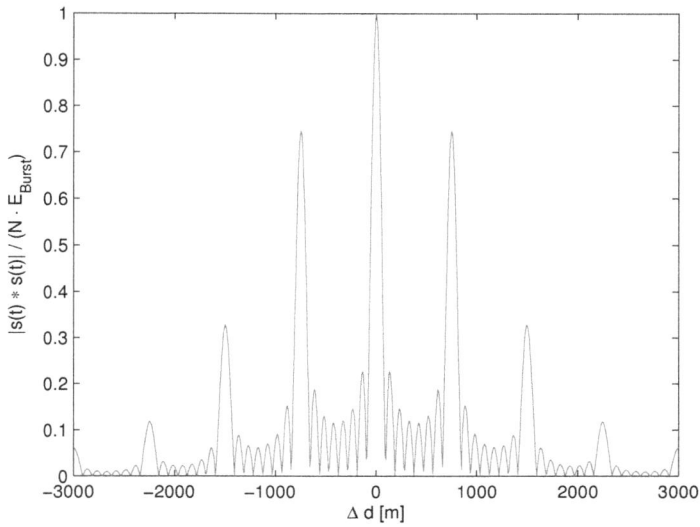

Figure 5.6: Absolute value of an ambiguity function with identical burst autocorrelation functions

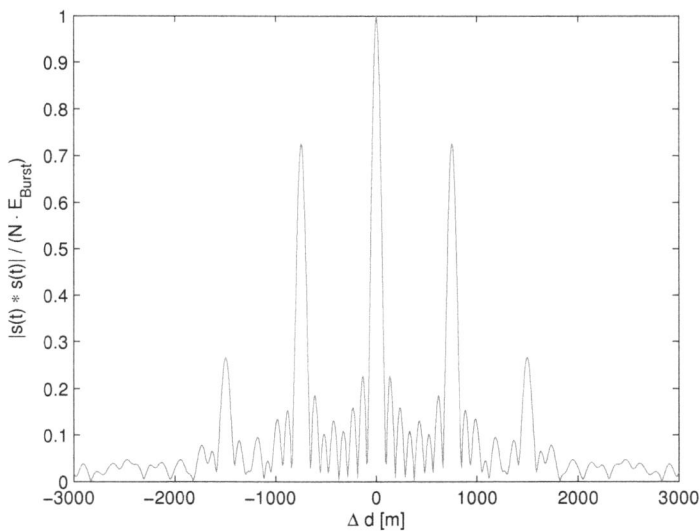

Figure 5.7: Absolute value of an ambiguity function with non-identical burst autocorrelation functions

5.1.5 Properties of the Ambiguity Function

In the following, the main properties of the ambiguity function are investigated. In order to obtain simple closed-form mathematical expressions, linear frequency sequences or permutations thereof are assumed. Similar results can be derived from [46] and [55].

Energy Considerations

The relationship between the energy of the correlated signals and the maximum value of the ambiguity function can be expressed as follows:

$$\left[s_{FH,ECB}(t) \star s_{FH,ECB}(t) \right]\Big|_{\Delta\tau=0} = E_{Total} = N E_{Burst} \tag{5.26}$$

This result is a direct consequence of the properties of autocorrelation functions [28] and radar ambiguity functions [54].

Therefore, the number of bursts N and the energy of the employed bursts E_{Burst} have a major influence on the best attainable signal to noise ratio and detectability of the main peak. Due to their increased length, the application of normal bursts provides an improvement compared to the application of access bursts.

Main Peak Width

The width of the main peak is a decisive characteristic of the ambiguity function since it determines the multipath separability of the estimation technique. The width corresponds to the resolution which solely depends on the bandwidth of the applied signal.

The width of the main peak (full width half maximum) results from an inverse Fourier transform of the energy density spectrum of the signal according to the Wiener–Khinchine theorem [56]. Therefore the width can be approximated as

$$\Delta d_{Mainpeak} \approx \frac{c_0}{B_{Hopping}} \tag{5.27}$$

with c_0 denoting the speed-of-light and $B_{Hopping}$ the hopping bandwidth of the signal. This value can be related to the resolution of the time difference of arrival estimation technique, i.e.

$$\Delta d_{Resolution} \approx \Delta d_{Mainpeak} \approx \frac{c_0}{B_{Hopping}} \tag{5.28}$$

As an example, an enlarged view on the main peak of an ambiguity function for $N = 8$ and $\Delta f_{Step} = 400\,\text{kHz}$ is shown in Fig. 5.8. In this case, the width of the main peak is approximately $100\,\text{m}$ which corresponds to the attained resolution.

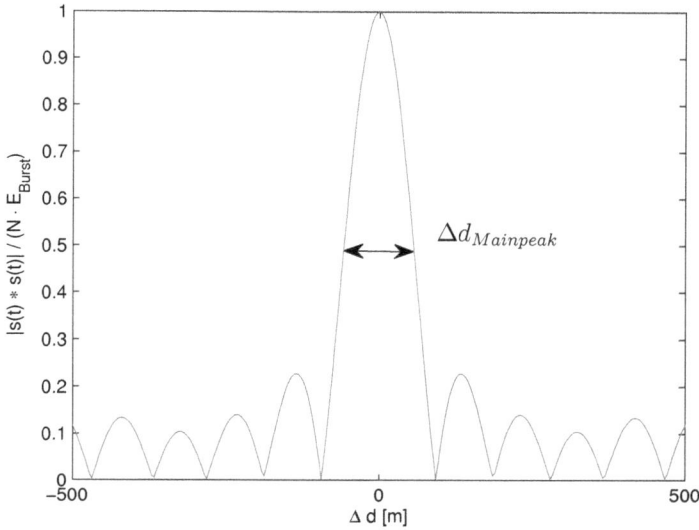

Figure 5.8: Enlarged view on the main peak of an ambiguity function

The main peak width also represents a bound on the bias error of the estimation technique in line-of-sight scenarios. This bias error can be approximated as

$$-\frac{c_0}{2B_{Hopping}} \leq \Delta\hat{d}_{BA} - \Delta d_{BA} \leq \frac{c_0}{2B_{Hopping}} \tag{5.29}$$

Therefore, a large hopping bandwidth is advantageous regarding the performance of the estimation technique.

Distance of Side Peaks

The shape of the ambiguity function is characterized by a main peak and side peaks at distinct distances from the main peak. These side peaks result from the structure of the multi-carrier response function which has been derived in a former section.

The multi-carrier response function exhibits a periodic structure with constructive superpositions at

$$\Delta d = m\frac{c_0}{\Delta f_{Step}} \quad \text{with} \quad m \in \mathbb{Z} \tag{5.30}$$

Therefore, the distance between the occurring peaks can be calculated as

$$\Delta d_{Main-Sidepeak} = \frac{c_0}{\Delta f_{Step}} \tag{5.31}$$

As an example, an ambiguity function for $N = 8$ and $\Delta f_{Step} = 400\,\mathrm{kHz}$ is shown in Fig. 5.9. In this case, the distance between the peaks is $750\,\mathrm{m}$.

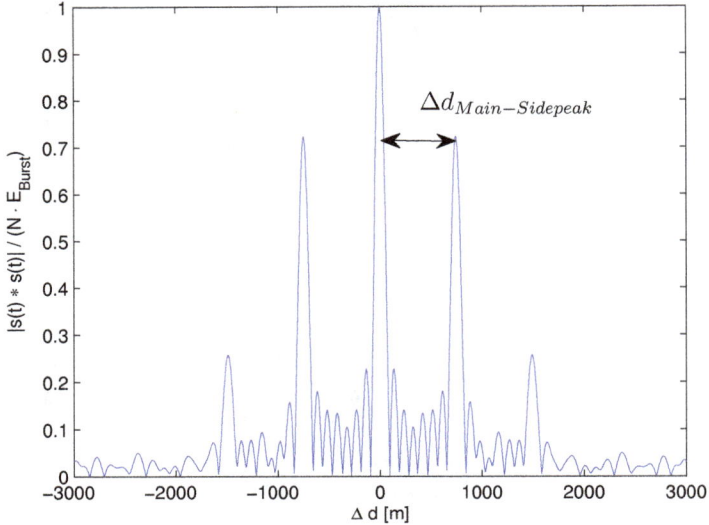

Figure 5.9: Distance of side peaks of an ambiguity function

The distance of the main peak to the side peaks has an impact on the maximum additive delay of a multipath component which may not interfere with the main peak.

Maximum Side Peak Attenuation

The attenuations of the side peaks result from the sampling of the autocorrelation functions of the narrowband burst signals and therefore only depend on the distance of the side peaks. Since the distance of the side peaks is determined by the frequency step size Δf_{Step}, the side peak attenuation is also solely dependent on this parameter.

The maximum side peak attenuation for different frequency step sizes Δf_{Step} is summarized in Tab. 5.1.

Δf_{Step}	200 kHz	400 kHz	600 kHz	800 kHz	1 MHz
$\Delta a_{Sidepeak}$	0.73	0.27	0.13	0.076	0.050

Table 5.1: Maximum side peak attenuation for different frequency step sizes

As an example, an ambiguity function for $N = 8$ and $\Delta f_{Step} = 400\,\text{kHz}$ is shown in Fig. 5.10. In this case, the maximum side peak attenuation is 0.27.

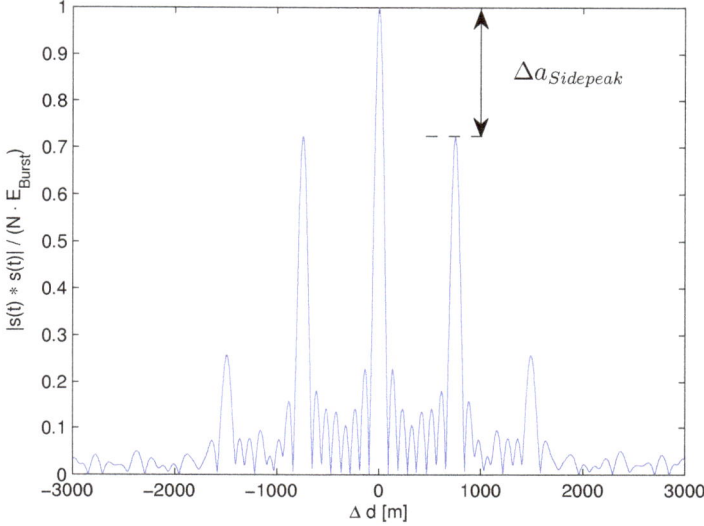

Figure 5.10: Maximum side peak attenuation of an ambiguity function

The maximum side peak attenuation has an impact on the range limit, since it influences the energy level of the main peak relative to the first side peak.

Maximum Side Lobe Attenuation

The ambiguity function exhibits side lobes around the main peak. The side lobes result from the band-limiting of the signal in the frequency domain, which can be interpreted as a rectangular window function.

The maximum side lobe attenuation for this window function is given as

$$\Delta a_{Sidelobe} \approx 0.78 \tag{5.32}$$

for any frequency hopping configuration of the signal. Similar results can be derived from the properties of ambiguity functions of radar signals. [54]

The maximum side lobe attenuation is constant and may influence the dynamic range of the estimation technique. As an example, the maximum side lobe attenuation for an ambiguity function with $N = 8$ and $\Delta f_{Step} = 400\,\text{kHz}$ is shown in Fig. 5.11.

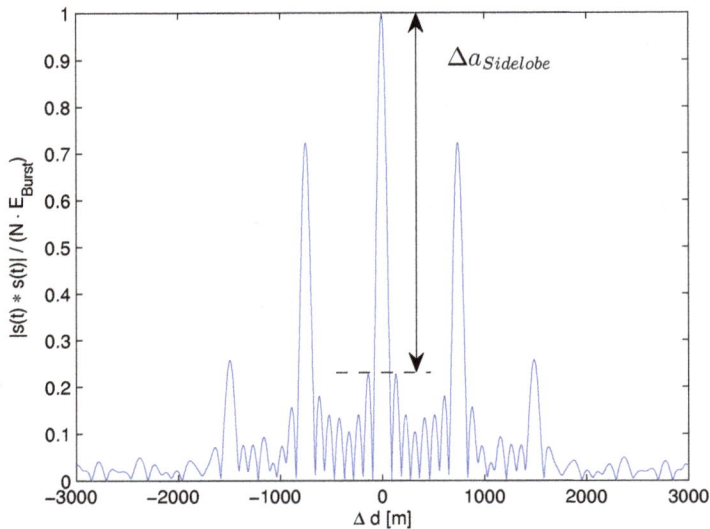

Figure 5.11: Maximum side lobe attenuation of an ambiguity function

5.1.6 Application of Concatenated and Stacked Signals

The acquisition of the wideband signal and the evaluation of the wideband crosscorrelation function requires a considerable amount of memory and processing power. Furthermore, the signal to noise ratio of the acquired wideband signals is substantially lower compared to the narrowband signals. In order to solve these challenges, the application of concatenated and stacked signals for crosscorrelation has been studied. This concept is presented in the following section.

The concatenation or stacking operation has to be performed identically in all receiving stations in order to ensure the applicability of the wideband crosscorrelation technique. If necessary, the relevant parameters have to be exchanged between the receiving stations beforehand.

Concatenated Signal Structure

The concatenated signal structure is characterized by assembling isolated burst signals one after the other. Consequently, the empty time slots, which are normally part of the acquired wideband signal, are omitted.

The model for the concatenated signal structure can be expressed as follows:

$$s_{Concatenated,ECB}(t) = \sum_{n=0}^{N-1} s_{Burst,BB}(t, \mathbf{b_n})e^{j(2\pi\Delta f_n t + \Delta\phi_n)} * \delta(t - nT_{Burst})$$

$$\text{with} \quad 0 \le t \le NT_{Burst} \tag{5.33}$$

In the sampled domain, the concatenated signal can be obtained as

$$s_{Concatenated,ECB}[k] = [s_{Burst,Isolated,0}[\cdot], s_{Burst,Isolated,1}[\cdot], \cdots, s_{Burst,Isolated,N-1}[\cdot]] \tag{5.34}$$

An exemplary spectrogram is depicted in Fig. 5.12.

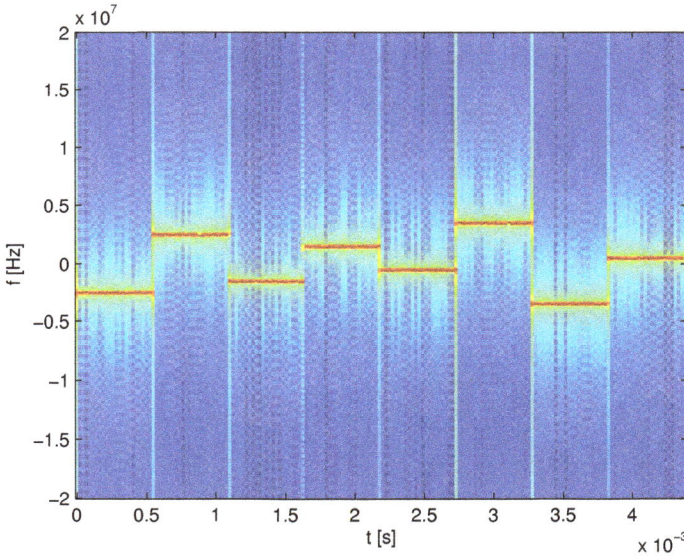

Figure 5.12: Concatenated signal structure in time and frequency

The overall signal to noise ratio $SNR_{Concatenated}$ can be calculated as

$$SNR_{Concatenated} = \frac{E_{Total}/T_{Total}}{\mathcal{N}_0 B_{Window}} = \frac{(NE_{Burst})/(NT_{Burst})}{\mathcal{N}_0 B_{Window}} \tag{5.35}$$

$$= \frac{E_{Burst}/T_{Burst}}{\mathcal{N}_0 B_{Narrowband}(B_{Window}/B_{Narrowband})}$$

$$= SNR_{Narrowband}\frac{B_{Narrowband}}{B_{Window}}$$

Stacked Signal Structure

The stacked signal structure is characterized by addition of all isolated burst signals. Consequently, the resulting signal contains multiple frequencies at the same time instants and can therefore be characterized as a multi-carrier signal.

The model for the stacked signal structure can be expressed as follows:

$$s_{Stacked,ECB}(t) = \sum_{n=0}^{N-1} s_{Burst,BB}(t, \mathbf{b_n}) e^{j(2\pi \Delta f_n t + \Delta \phi_n)} \tag{5.36}$$

$$\text{with} \quad 0 \leq t \leq T_{Burst}$$

In the sampled domain, the stacked signal can be obtained as

$$s_{Stacked,ECB}[k] = \sum_{n=0}^{N-1} s_{Burst,Isolated,n}[\cdot] \tag{5.37}$$

An exemplary spectrogram is shown in Fig. 5.13.

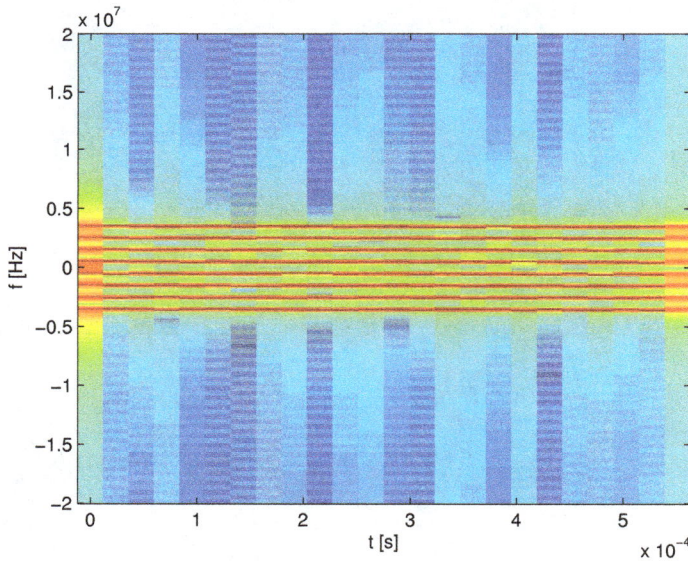

Figure 5.13: Stacked signal structure in time and frequency

The overall signal to noise ratio $SNR_{Stacked}$ can be calculated as

$$SNR_{Stacked} = \frac{E_{Total}/T_{Total}}{N\mathcal{N}_0 B_{Window}} = \frac{(NE_{Burst})/T_{Burst}}{N\mathcal{N}_0 B_{Window}} \tag{5.38}$$

$$= \frac{E_{Burst}/T_{Burst}}{\mathcal{N}_0 B_{Narrowband}(B_{Window}/B_{Narrowband})}$$

$$= SNR_{Narrowband}\frac{B_{Narrowband}}{B_{Window}}$$

Crosscorrelation Properties

The crosscorrelation properties are mainly dependent on the shape of the ambiguity function of the signal. In the former sections, the ambiguity function for the complete frequency hopping signal $s_{FH,ECB}(t) \star s_{FH,ECB}(t)$ has been derived and analyzed.

The ambiguity function of the concatenated and stacked signal can be obtained by replacing $s_{FH,ECB}(t)$ by $s_{Concatenated,ECB}(t)$ or $s_{Stacked,ECB}(t)$, respectively. This approach leads to the same intermediate expression

$$\sum_{n=0}^{N-1}\left(s_{Burst,BB}(t,\mathbf{b_n})e^{j(2\pi\Delta f_n t+\Delta\phi_n)}\right) \star \left(s_{Burst,BB}(t,\mathbf{b_n})e^{j(2\pi\Delta f_n t+\Delta\phi_n)}\right) \tag{5.39}$$

which has been obtained for the complete frequency hopping signal. Consequently, all subsequent derivations are essentially the same and the crosscorrelation properties remain unaffected.[1]

Signal to Noise Ratio Enhancement

The signal to noise ratio of the acquired wideband signal is much lower compared to the case of narrowband signals due to the empty time slots and the large acquisition bandwidth of the frontend. During the pre-processing, the narrowband burst signals are isolated and the noise in time and frequency is removed.

The signal to noise ratio of the concatenated and stacked signal can be summarized as

$$SNR_{Concatenated} = SNR_{Stacked} = SNR_{Narrowband}\frac{B_{Narrowband}}{B_{Window}} \tag{5.40}$$

This expression only depends on the bandwidth of the filtering window B_{Window} which has been applied during the burst isolation process. The signal to noise ration of the narrowband signal is the best attainable case and can be reached by setting the filter window bandwidth to $B_{Window} = B_{Narrowband} = 200\,\mathrm{kHz}$. Thus, the low signal to noise ratio of acquired wideband signals can be improved significantly.

---

[1]Since the concatenated and stacked signals are shorter in time, the domain of definition for time delay in the crosscorrelation function is smaller but still adequate for typical scenarios.

Data Reduction and Processing Speed

The wideband signal acquisition and sampling of the entire GSM uplink band over several TDMA frames requires a considerable amount of memory. Assuming a quantization of b bit, the required memory amounts to

$$\text{Memory} = N f_s T_{Frame}\, b/4 \text{ Byte} \tag{5.41}$$

Choosing a sampling rate of $f_s = 40\,\text{MHz}$ (covering the E-GSM 900 uplink band of $35\,\text{MHz}$ bandwidth) and a quantization of $b = 10\,\text{Bit}$ yields

$$\text{Memory} = N 461.5\,\text{kByte} \tag{5.42}$$

The algorithms for time difference of arrival estimation require that all received signals are available at a central processing unit and therefore the signals have to be collected and transmitted over a data link. Furthermore, the application of the algorithms on large data files takes a significant amount of time.

A considerable advantage of the concatenation and stacking technique is the fact that they can be carried out on the receiving stations locally. Thus, a substantially smaller amount of data has to be collected and transmitted to the central processing unit. In case of a concatenated signal, the amount of data is reduced by a factor of approximately 8 for normal bursts and 16 for access bursts. When using a stacked signal, a reduction of a factor of approximately $8N$ for normal bursts and $16N$ for access bursts is possible.

5.1.7 Range Limitations

The existence of side peaks in the ambiguity function may introduce a limitation of the applicable range of the localization system.

In a simplified case with one additional multipath component in one radio channel impulse response, the combined impulse response can be expressed as

$$h_{A,ECB}(t) \star h_{B,ECB}(t) = \delta(t) + \underline{\alpha}_{MP}\delta(t - \tau_{MP}) \tag{5.43}$$

with $\underline{\alpha}_{MP}$ denoting the relative complex amplitude of the multipath component and τ_{MP} its relative time delay with respect to the direct path component.

In the case of a multipath component traveling at a relative time delay equal to the side peak distance, i.e.

$$c_0\tau_{MP} = \Delta d_{Main-Sidepeak} = \frac{c_0}{\Delta f_{Step}} \tag{5.44}$$

the main peak of the direct path ambiguity function interferes with the first side peak of the multipath ambiguity function. In this case, the wideband crosscorrelation function can be severely distorted and the maximum value may not provide a reliable estimate for the time difference of arrival. A solution to this challenge is the requirement, that the multipath component is sufficiently attenuated relative to the direct path component.

The inherent attenuation of the first side peak relative to the main peak can be taken into account using the factor $(1 - \Delta a_{Sidepeak})$. The attenuation due to radio signal propagation can be approximated using a *Free Space Path Loss (FSPL)* model. This model describes the attenuation of electromagnetic waves under the assumption of lossless isotropic antennas and can be expressed as

$$\text{FSPL}(d) = \left(\frac{4\pi d f_c}{c_0}\right)^2 \tag{5.45}$$

with d denoting the absolute distance between the transmitter and the receiver. [34]

Consequently, the requirement for the relative attenuations can be stated as

$$(1 - \Delta a_{Sidepeak}) \frac{\text{FSPL}(d)}{\text{FSPL}(d + c_0/\Delta f_{Step})} \overset{!}{\leq} \text{ratio} \tag{5.46}$$

In this expression, the term ratio denotes the desired relative attenuation of the multipath component compared to the direct path component. Using the free space path loss model, this expression can be reduced to

$$d \leq \frac{c_0}{\Delta f_{Step}} \frac{\sqrt{\text{ratio}/(1 - \Delta a_{Sidepeak})}}{1 - \sqrt{\text{ratio}/(1 - \Delta a_{Sidepeak})}} \tag{5.47}$$

yielding a range condition for the maximum absolute distance between the signal source and the receiving stations.

The results for the maximum absolute distance d_{Max} for different attenuation ratios of $1/10$ and $1/100$ are summarized in Tab. 5.2 and Tab. 5.3, respectively.

Δf_{Step}	200 kHz	400 kHz	600 kHz	800 kHz	1 MHz
$\Delta a_{Sidepeak}$	0.73	0.27	0.13	0.076	0.050
d_{Max}	2332.2 m	440.7 m	256.5 m	183.8 m	144.1 m

Table 5.2: Maximum distances for an attenuation ratio of $1/10$

Δf_{Step}	200 kHz	400 kHz	600 kHz	800 kHz	1 MHz
$\Delta a_{Sidepeak}$	0.73	0.27	0.13	0.076	0.050
d_{Max}	357.5 m	99.4 m	60.0 m	43.5 m	34.3 m

Table 5.3: Maximum distances for an attenuation ratio of $1/100$

Therefore, limiting the area of operation to smaller sizes than the maximum absolute distance enables a reliable application of the wideband crosscorrelation technique.

5.2 Burst Phase Analysis Technique

The second coherent technique for time difference of arrival estimation is based on the analysis of burst phase values and is described in the following sections.

5.2.1 Derivation for Simple Delay Scenarios

The technique is based on the analysis of phase values of burst signals emitted at different carrier frequencies. Therefore several burst signals have to be acquired by the receiving stations.

The following derivations are based on the wideband signal acquisition models. The derivation for narrowband signal acquisition can be carried out accordingly yielding the same results.

For simplicity, the radio channels are modeled as simple delay channels, i.e.

$$h_{A,ECB}(t) = \underline{\alpha}\delta(t - \tau_A)e^{-j2\pi f_{LO}\tau_A} \tag{5.48}$$

$$h_{B,ECB}(t) = \underline{\beta}\delta(t - \tau_B)e^{-j2\pi f_{LO}\tau_B} \tag{5.49}$$

with $\underline{\alpha}$ and $\underline{\beta}$ denoting the complex amplitudes, τ_A and τ_B representing the respective time delays and f_{LO} denoting the corresponding local oscillator frequency.

Assuming that different burst signals $s_{Burst,ECB}(t, \mathbf{b_n})$ with modulating bit sequences $\mathbf{b_n}$ are transmitted by the mobile station and that noise terms are negligible, the received signals for burst n can then be expressed as

$$r_{A,ECB,n}(t) = \underline{\alpha}s_{Burst,ECB}(t - \tau_A, \mathbf{b_n})e^{-j2\pi f_{LO}\tau_A} \tag{5.50}$$

$$r_{B,ECB,n}(t) = \underline{\beta}s_{Burst,ECB}(t - \tau_B, \mathbf{b_n})e^{-j2\pi f_{LO}\tau_B} \tag{5.51}$$

As a first step, the quotient between the two received signals is calculated as

$$q_n(t) = \frac{r_{B,ECB,n}(t)}{r_{A,ECB,n}(t)} = \frac{\underline{\beta}s_{Burst,ECB}(t - \tau_B, \mathbf{b_n})}{\underline{\alpha}s_{Burst,ECB}(t - \tau_A, \mathbf{b_n})}e^{-j2\pi f_{LO}(\tau_B - \tau_A)} \tag{5.52}$$

Using the relationship between the acquired burst signals and their baseband representation for wideband signal acquisition

$$s_{Burst,ECB}(t, \mathbf{b_n}) = s_{Burst,BB}(t, \mathbf{b_n})e^{j(2\pi\Delta f_n t + \Delta\phi_n)} \tag{5.53}$$

the quotient can be simplified to

$$
\begin{aligned}
q_n(t) &= \frac{\underline{\beta}s_{Burst,BB}(t - \tau_B, \mathbf{b_n})}{\underline{\alpha}s_{Burst,BB}(t - \tau_A, \mathbf{b_n})}e^{-j2\pi\Delta f_n(\tau_B - \tau_A)}e^{-j2\pi f_{LO}(\tau_B - \tau_A)} \\
&= \frac{\underline{\beta}s_{Burst,BB}(t - \tau_B, \mathbf{b_n})}{\underline{\alpha}s_{Burst,BB}(t - \tau_A, \mathbf{b_n})}e^{-j2\pi f_{c,n}(\tau_B - \tau_A)}
\end{aligned}
\tag{5.54}
$$

Since the modulation of the burst signal is a pure phase modulation, i.e.

$$s_{Burst,BB}(t, \mathbf{b_n}) = e^{j\phi(t,\mathbf{b_n})} \tag{5.55}$$

the quotient can be further reduced to

$$q_n(t) = \frac{\beta}{\alpha} e^{j(\phi(t-\tau_B,\mathbf{b_n}) - \phi(t-\tau_A,\mathbf{b_n}))} e^{-j2\pi f_{c,n}(\tau_B - \tau_A)} \tag{5.56}$$

The value of the phase is influenced by the complex amplitudes of the radio channels, the time-dependent burst phase differences and the phase shift caused by the delay of the carrier signal. The phase value of an exemplary quotient $q_n(t)$ focused on the training sequence of a normal burst and its mean value are depicted in Fig. 5.14.

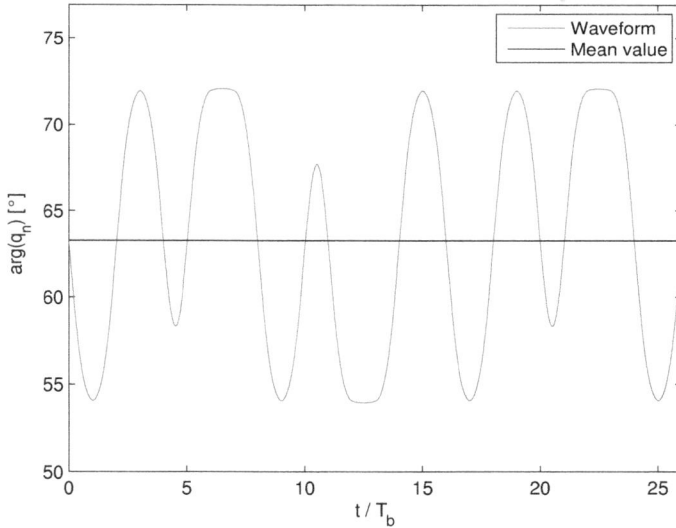

Figure 5.14: Phase value and mean value of a quotient $q_n(t)$

The complex amplitudes of the radio channels and the phase shift of the carrier signal are constant values over the length of the burst signal. The time-dependency is introduced by the GMSK modulation and varies around a constant (but still unknown) phase value.

For the application of the technique, this time-dependency is undesired and it is assumed, that the time-varying phase can be replaced by a constant phase term $\Delta\phi_{GMSK}$ based on the mean value over the training or synchronization sequence. The validity of this assumption is verified in a forthcoming section.

The resulting mean quotient can then be expressed as

$$\bar{q}_n = \frac{\beta}{\alpha} e^{j\Delta\phi_{GMSK}} e^{-j2\pi f_{c,n}(\tau_B - \tau_A)} \tag{5.57}$$

The next challenge arises from the carrier phase shift which causes range ambiguities in the scale of $c_0/f_{c,n}$ which is approximately 30 cm in the case of E-GSM 900 systems. Furthermore, the complex amplitudes of the radio channels and the mean phase value $\Delta\phi_{GMSK}$ are still unknown. An interesting solution is the application of a frequency hopping scheme between consecutive burst signals. For simplicity, it is assumed that two burst signals are transmitted on different carrier frequencies given by

$$f_{c,n+1} = f_{c,n} + \Delta f_{Step} \tag{5.58}$$

The frequency step size between the carrier frequencies is denoted as Δf_{Step}. Consequently, a further quotient may be defined as

$$
\begin{aligned}
\frac{\bar{q}_{n+1}}{\bar{q}_n} &= \frac{\beta/\alpha \; e^{j\Delta\phi_{GMSK}} e^{-j2\pi f_{c,n+1}(\tau_B - \tau_A)}}{\beta/\alpha \; e^{j\Delta\phi_{GMSK}} e^{-j2\pi f_{c,n}(\tau_B - \tau_A)}} \\
&= \frac{e^{-j2\pi(f_{c,n} + \Delta f_{Step})(\tau_B - \tau_A)}}{e^{-j2\pi f_{c,n}(\tau_B - \tau_A)}} \\
&= e^{-j2\pi\Delta f_{Step}(\tau_B - \tau_A)}
\end{aligned}
\tag{5.59}
$$

This expression is independent of the unknown complex amplitudes of the radio channels and the mean phase value $\Delta\phi_{GMSK}$. Furthermore, the carrier frequencies $f_{c,n}$ are no longer present in the expression and are replaced by the frequency step size Δf_{Step}. For E-GSM 900 systems, the minimum frequency step size Δf_{Step} is 200 kHz which leads to range ambiguities in the scale of 1.5 km. For many interesting applications, this unambiguous range poses no restrictions on the applicability of the technique.

The time difference of arrival can finally be estimated as

$$\Delta\hat{\tau}_{BA} = \tau_B - \tau_A = \frac{\arg(\bar{q}_{n+1}/\bar{q}_n)}{-2\pi\Delta f_{Step}} \tag{5.60}$$

5.2.2 Evaluation of Multiple Burst Signals

The derived expression is based on two bursts with different carrier frequencies. In this section, an expansion to multiple burst signals is introduced.

Considering the further quotient of the upper derivations, a recursion formula can be recognized. Consequently, consecutive mean quotients can be characterized by a constant phase decrement of $2\pi\Delta f_{Step}(\tau_B - \tau_A)$. Therefore, a series can be defined as

$$\bar{q}_n = e^{-j2\pi\Delta f_{Step} n(\tau_B - \tau_A)} e^{j\phi_{Series,0}} \tag{5.61}$$

with $\phi_{Series,0}$ representing the initial phase value of the series. This definition requires that each carrier frequency has a constant offset relative to the previous carrier frequency. Therefore, the carrier frequencies have to increase or decrease linearly.

For the evaluation of this sequence and estimation of the time difference of arrival $\Delta\tau_{BA} = \tau_B - \tau_A$, two different approaches may be applied as described in [47].

Least Squares Approximation

The first approach is based on determining the phase values of the series elements and calculating a linear *Least Squares (LS)* approximation. The slope of the straight line corresponds to the constant phase increment and can be used to estimate the time difference of arrival $\Delta\tau_{BA}$. An example is given in Fig. 5.15.

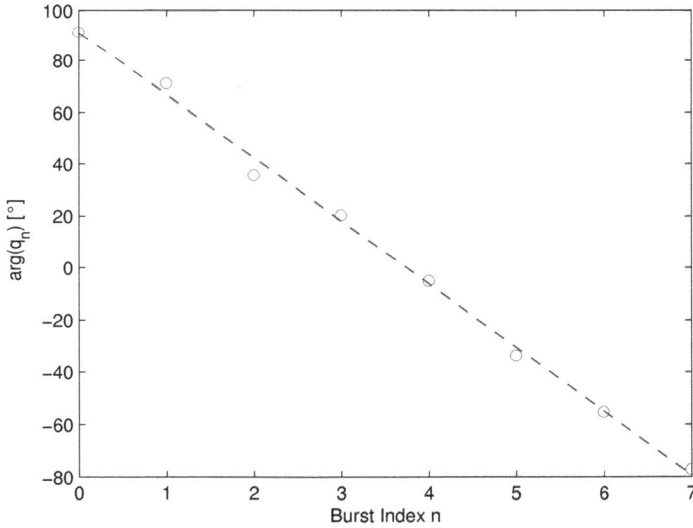

Figure 5.15: Phase value sequence with linear least squares approximation

The mathematical relationship can be expressed as

$$-2\pi\Delta f_{Step}(\tau_B - \tau_A) \overset{!}{=} \frac{\sum\limits_{n=0}^{N-1} n\arg(\overline{q}_n) - 1/N \sum\limits_{n=0}^{N-1} n \sum\limits_{n=0}^{N-1} \arg\overline{q}_n}{\sum\limits_{n=0}^{N-1} n^2 - 1/N (\sum\limits_{n=0}^{N-1} n)^2} \tag{5.62}$$

using the standard definition of the linear least squares approximation [57].

Since the phase values are 2π-periodic, a phase unwrapping may be necessary before calculating the least squares approximation. The time difference of arrival can then be estimated as

$$\Delta\hat{\tau}_{BA} = \frac{1}{-2\pi\Delta f_{Step}} \frac{\sum\limits_{n=0}^{N-1} n\arg(\bar{q}_n) - 1/N \sum\limits_{n=0}^{N-1} n \sum\limits_{n=0}^{N-1} \arg\bar{q}_n}{\sum\limits_{n=0}^{N-1} n^2 - 1/N(\sum\limits_{n=0}^{N-1} n)^2} \tag{5.63}$$

Inverse Discrete Fourier Transform

An alternative approach is based on an *Inverse Discrete Fourier Transform (IDFT)* of the series. The series may be interpreted as a complex exponential series with a normalized angular frequency of $-2\pi\Delta f_{Step}(\tau_B - \tau_A)$. The absolute value of an exemplary inverse Fourier transform is depicted in Fig. 5.16.

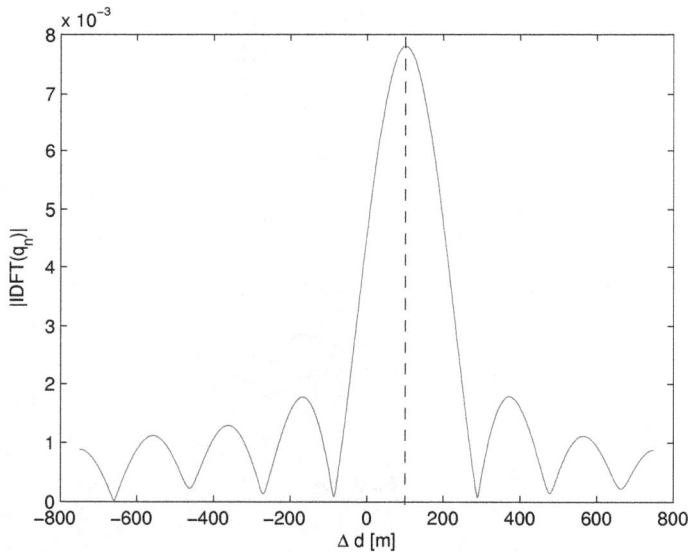

Figure 5.16: Absolute value of an inverse discrete Fourier transform

Calculating the inverse discrete Fourier transform and searching for the maximum in the absolute value then yields a normalized angular frequency which corresponds to the estimated time difference of arrival. Mathematically, this relationship can be expressed as

$$\bar{Q}_k = \text{IDFT}\{\bar{q}_n\} = \frac{1}{N} \sum\limits_{n=0}^{N-1} \bar{q}_n e^{j\frac{2\pi}{N}kn} \quad \text{with } 0 \leq k \leq N-1 \tag{5.64}$$

using the standard definition of the inverse discrete Fourier transform [28]. This expression can be simplified as

$$\overline{Q}_k = \frac{1}{N} \sum_{n=0}^{N-1} e^{-j2\pi\Delta f_{Step}n(\tau_B-\tau_A)} e^{j\phi_{Series,0}} e^{j\frac{2\pi}{N}kn} \tag{5.65}$$

$$= e^{j\phi_{Series,0}} \frac{1}{N} \sum_{n=0}^{N-1} e^{-j2\pi\Delta f_{Step}n(\tau_B-\tau_A)} e^{j\frac{2\pi}{N}kn}$$

$$= e^{j\phi_{Series,0}} \frac{1}{N} \sum_{n=0}^{N-1} e^{-j2\pi n(\Delta f_{Step}(\tau_B-\tau_A)-k/N)}$$

For increasing the accuracy, the inverse Fourier transform may be interpolated using zero-padding of the initial complex exponential series. Consequently, a quasi-continuous evaluation is possible. Using this assumption, the maximum absolute value is obtained for

$$\tilde{k}_{Max} \overset{!}{=} \tilde{N}\Delta f_{Step}(\tau_B - \tau_A) \tag{5.66}$$

with $\tilde{k}_{Max}$ denoting the index of the maximum and $\tilde{N}$ the length of the interpolated inverse Fourier transform.

For the evaluation of time difference of arrival estimates, the periodicity of the inverse Fourier transform has to be taken in account.

Positive estimates $\tau_B - \tau_A > 0$ are located in the first half of the inverse Fourier transform sequence and can be expressed as

$$\Delta\hat{\tau}_{BA} = \frac{\tilde{k}_{Max}/\tilde{N}}{\Delta f_{Step}} \quad \text{for} \quad 0 \leq \tilde{k}_{Max} < \tilde{N}/2 \tag{5.67}$$

Negative estimates $\tau_B - \tau_A < 0$ are situated in the second half of the inverse Fourier transform sequence and can be calculated as

$$\Delta\hat{\tau}_{BA} = -\frac{(\tilde{N} - \tilde{k}_{Max})/\tilde{N}}{\Delta f_{Step}} \quad \text{for} \quad \tilde{N}/2 \leq \tilde{k}_{Max} < \tilde{N} \tag{5.68}$$

5.2.3 Influence of Multipath Components

In this section, the influence of multipath components on the estimated time difference of arrival is investigated. The derivations are based on the same assumptions and requirements as in the previous sections with extensions to a more complex radio channel model.

The impulse responses of the radio channels are modeled as a superposition of weighted and shifted delta functions according

$$h_{A,ECB}(t) = \sum_{i=0}^{I-1} \underline{\alpha}_i \delta(t - \tau_{A,i}) e^{-j2\pi f_{LO}\tau_{A,i}} \tag{5.69}$$

$$h_{B,ECB}(t) = \sum_{j=0}^{J-1} \underline{\beta}_j \delta(t - \tau_{B,j}) e^{-j2\pi f_{LO}\tau_{B,j}} \tag{5.70}$$

with $\underline{\alpha}_i$ and $\underline{\beta}_j$ denoting the complex amplitudes, $\tau_{A,i}$ and $\tau_{B,j}$ the respective time delays and I and J the corresponding number of multipath components.

The received signals for burst n can then be expressed as

$$r_{A,ECB,n}(t) = \sum_{i=0}^{I-1} \underline{\alpha}_i s_{Burst,ECB}(t - \tau_{A,i}, \mathbf{b_n}) e^{-j2\pi f_{LO}\tau_{A,i}} \tag{5.71}$$

$$r_{B,ECB,n}(t) = \sum_{j=0}^{J-1} \underline{\beta}_j s_{Burst,ECB}(t - \tau_{B,j}, \mathbf{b_n}) e^{-j2\pi f_{LO}\tau_{B,j}} \tag{5.72}$$

Using these models for the received signals, the quotient between the signals can be stated as

$$q_n(t) = \frac{r_{B,ECB,n}(t)}{r_{A,ECB,n}(t)} = \frac{\sum\limits_{j=0}^{J-1} \underline{\beta}_j s_{Burst,ECB}(t - \tau_{B,j}, \mathbf{b_n}) e^{-j2\pi f_{LO}\tau_{B,j}}}{\sum\limits_{i=0}^{I-1} \underline{\alpha}_i s_{Burst,ECB}(t - \tau_{A,i}, \mathbf{b_n}) e^{-j2\pi f_{LO}\tau_{A,i}}} \tag{5.73}$$

Applying the relationship between the acquired burst signals and their baseband representation for wideband signal acquisition

$$s_{Burst,ECB}(t, \mathbf{b_n}) = s_{Burst,BB}(t, \mathbf{b_n}) e^{j(2\pi\Delta f_n t + \Delta\phi_n)} \tag{5.74}$$

the quotient between the received signals can be expressed as

$$q_n(t) = \frac{\sum\limits_{j=0}^{J-1} \underline{\beta}_j s_{Burst,BB}(t - \tau_{B,j}, \mathbf{b_n}) e^{-j2\pi\Delta f_n \tau_{B,j}} e^{-j2\pi f_{LO}\tau_{B,j}}}{\sum\limits_{i=0}^{I-1} \underline{\alpha}_i s_{Burst,BB}(t - \tau_{A,i}, \mathbf{b_n}) e^{-j2\pi\Delta f_n \tau_{A,i}} e^{-j2\pi f_{LO}\tau_{A,i}}}$$

$$= \frac{\sum\limits_{j=0}^{J-1} \underline{\beta}_j s_{Burst,BB}(t - \tau_{B,j}, \mathbf{b_n}) e^{-j2\pi f_{c,n}\tau_{B,j}}}{\sum\limits_{i=0}^{I-1} \underline{\alpha}_i s_{Burst,BB}(t - \tau_{A,i}, \mathbf{b_n}) e^{-j2\pi f_{c,n}\tau_{A,i}}} \tag{5.75}$$

Using the phase modulation model for the baseband signal according

$$s_{Burst,BB}(t, \mathbf{b_n}) = e^{j\phi(t,\mathbf{b_n})} \tag{5.76}$$

the quotient can finally be expressed as

$$q_n(t) = \frac{\displaystyle\sum_{j=0}^{J-1} \underline{\beta}_j e^{j\phi(t-\tau_{B,j},\mathbf{b_n})} e^{-j2\pi f_{c,n}\tau_{B,j}}}{\displaystyle\sum_{i=0}^{I-1} \underline{\alpha}_i e^{j\phi(t-\tau_{A,i},\mathbf{b_n})} e^{-j2\pi f_{c,n}\tau_{A,i}}} \tag{5.77}$$

This expression shows many similarities to the simple delay scenario. Due to the coherent summation in the numerator and denominator, a direct evaluation of the time difference of arrival between the received signals is not possible. In order to enable further simplifications, the quotient is expanded by the complex conjugate of the denominator yielding

$$q_n(t) = \frac{\left(\displaystyle\sum_{j=0}^{J-1} \underline{\beta}_j e^{j\phi(t-\tau_{B,j},\mathbf{b_n})} e^{-j2\pi f_{c,n}\tau_{B,j}}\right)\left(\displaystyle\sum_{i=0}^{I-1} \underline{\alpha}_i^* e^{-j\phi(t-\tau_{A,i},\mathbf{b_n})} e^{j2\pi f_{c,n}\tau_{A,i}}\right)}{\left(\displaystyle\sum_{i=0}^{I-1} \underline{\alpha}_i e^{j\phi(t-\tau_{A,i},\mathbf{b_n})} e^{-j2\pi f_{c,n}\tau_{A,i}}\right)\left(\displaystyle\sum_{i=0}^{I-1} \underline{\alpha}_i^* e^{-j\phi(t-\tau_{A,i},\mathbf{b_n})} e^{j2\pi f_{c,n}\tau_{A,i}}\right)}$$

$$= \frac{\displaystyle\sum_{i=0}^{I-1}\sum_{j=0}^{J-1} \underline{\alpha}_i^* \underline{\beta}_j e^{j(\phi(t-\tau_{B,j},\mathbf{b_n})-\phi(t-\tau_{A,i},\mathbf{b_n}))} e^{-j2\pi f_{c,n}(\tau_{B,j}-\tau_{A,i})}}{\left|\displaystyle\sum_{i=0}^{I-1} \underline{\alpha}_i e^{j\phi(t-\tau_{A,i},\mathbf{b_n})} e^{-j2\pi f_{c,n}\tau_{A,i}}\right|^2} \tag{5.78}$$

In order to obtain time-independent expressions in the numerator, the time-dependent burst phase differences are replaced by constant phase terms $\Delta\phi_{GMSK,i,j}$ based on the respective mean values over the training or synchronization sequence. The validity of this assumption is verified in a forthcoming section.

Consequently, the quotient can be expressed as

$$\bar{q}_n = \frac{\displaystyle\sum_{i=0}^{I-1}\sum_{j=0}^{J-1} \underline{\alpha}_i^* \underline{\beta}_j e^{j\Delta\phi_{GMSK,i,j}} e^{-j2\pi f_{c,n}(\tau_{B,j}-\tau_{A,i})}}{\left|\displaystyle\sum_{i=0}^{I-1} \underline{\alpha}_i e^{j\phi(t-\tau_{A,i},\mathbf{b_n})} e^{-j2\pi f_{c,n}\tau_{A,i}}\right|^2} \tag{5.79}$$

In order to investigate the influence of multipath components on the phase increment between the quotients of two consecutive burst signals, it is now assumed that

$$f_{c,n+1} = f_{c,n} + \Delta f_{Step} \tag{5.80}$$

The further quotient between consecutive burst signals can then be expressed as

$$
\frac{\overline{q}_{n+1}}{\overline{q}_n} = \frac{\displaystyle\sum_{i=0}^{I-1}\sum_{j=0}^{J-1} \underline{\alpha}_i^* \underline{\beta}_j e^{j\Delta\phi_{GMSK,i,j}} e^{-j2\pi(f_{c,n}+\Delta f_{Step})(\tau_{B,j}-\tau_{A,i})}}{\displaystyle\sum_{i=0}^{I-1}\sum_{j=0}^{J-1} \underline{\alpha}_i^* \underline{\beta}_j e^{j\Delta\phi_{GMSK,i,j}} e^{-j2\pi f_{c,n}(\tau_{B,j}-\tau_{A,i})}}
$$

$$
\cdot \frac{\left| \displaystyle\sum_{i=0}^{I-1} \underline{\alpha}_i e^{j\phi(t-\tau_{A,i},\mathbf{b_n})} e^{-j2\pi f_{c,n}\tau_{A,i}} \right|^2}{\left| \displaystyle\sum_{i=0}^{I-1} \underline{\alpha}_i e^{j\phi(t-\tau_{A,i},\mathbf{b_{n+1}})} e^{-j2\pi(f_{c,n}+\Delta f_{Step})\tau_{A,i}} \right|^2} \tag{5.81}
$$

Assuming that the components $\tau_{A,0}$ and $\tau_{B,0}$ correspond to the desired line-of-sight components, this expression can be rearranged as

$$
\frac{\overline{q}_{n+1}}{\overline{q}_n} = \frac{\displaystyle\sum_{i=0}^{I-1}\sum_{j=0}^{J-1} \underline{\alpha}_i^* \underline{\beta}_j e^{j\Delta\phi_{GMSK,i,j}} e^{-j2\pi f_{c,n}(\tau_{B,j}-\tau_{A,i})} e^{-j2\pi\Delta f_{Step}(\tau_{B,j}-\tau_{A,i})}}{\displaystyle\sum_{i=0}^{I-1}\sum_{j=0}^{J-1} \underline{\alpha}_i^* \underline{\beta}_j e^{j\Delta\phi_{GMSK,i,j}} e^{-j2\pi f_{c,n}(\tau_{B,j}-\tau_{A,i})} e^{-j2\pi\Delta f_{Step}(\tau_{B,0}-\tau_{A,0})}}
$$

$$
\cdot \frac{\left| \displaystyle\sum_{i=0}^{I-1} \underline{\alpha}_i e^{j\phi(t-\tau_{A,i},\mathbf{b_n})} e^{-j2\pi f_{c,n}\tau_{A,i}} \right|^2}{\left| \displaystyle\sum_{i=0}^{I-1} \underline{\alpha}_i e^{j\phi(t-\tau_{A,i},\mathbf{b_{n+1}})} e^{-j2\pi f_{c,n}\tau_{A,i}} e^{-j2\pi\Delta f_{Step}\tau_{A,i}} \right|^2} e^{-j2\pi\Delta f_{Step}(\tau_{B,0}-\tau_{A,0})} \tag{5.82}
$$

Due to the multipath components, the amplitude and phase of the desired exponential function are influenced. Since the estimation technique relies on the evaluation of the phase values, only the first factor has an impact on the estimation performance. The induced phase deviation can be expressed as

$$
\Delta\phi_{Deviation} = \arg\left(\frac{\displaystyle\sum_{i=0}^{I-1}\sum_{j=0}^{J-1} \underline{\alpha}_i^* \underline{\beta}_j e^{j\Delta\phi_{GMSK,i,j}} e^{-j2\pi f_{c,n}(\tau_{B,j}-\tau_{A,i})} e^{-j2\pi\Delta f_{Step}(\tau_{B,j}-\tau_{A,i})}}{\displaystyle\sum_{i=0}^{I-1}\sum_{j=0}^{J-1} \underline{\alpha}_i^* \underline{\beta}_j e^{j\Delta\phi_{GMSK,i,j}} e^{-j2\pi f_{c,n}(\tau_{B,j}-\tau_{A,i})} e^{-j2\pi\Delta f_{Step}(\tau_{B,0}-\tau_{A,0})}} \right)
$$

$$\tag{5.83}$$

The phase deviation can then be related to the deviation of the time difference of arrival estimate as

$$
\Delta\hat{\tau}_{BA} - \Delta\tau_{BA} = \frac{\Delta\phi_{Deviation}}{-2\pi\Delta f_{Step}} \tag{5.84}
$$

The phase deviation mainly depends on the complex amplitudes $\underline{\alpha}_i$ and $\underline{\beta}_j$ as well as the corresponding time delays $\tau_{A,i}$ and $\tau_{B,j}$ of the radio channels. Furthermore, the burst phase differences $\Delta\phi_{GMSK,i,j}$ and the carrier frequencies $f_{c,n}$ have to be considered.

In order to gain a further understanding of the estimation deviations, the complex quotient is investigated in detail.

The numerator and the denominator can be interpreted as a sum of vectors in the complex plane. The first summands with indices $i = j = 0$ are equal in absolute value and phase value and therefore represent identical vectors. Other corresponding summands with same i and j have equal absolute value and only differ in the phase value. The resulting vectors of the numerator and denominator can be obtained by vector summation of the constituting vectors. An arbitrary scenario with one multipath component in each radio channel impulse response is depicted in Fig. 5.17. The two dashed lines represent the resulting vectors of the numerator and denominator. The angle between these two vectors can be identified as the phase deviation $\Delta\phi_{Deviation}$.

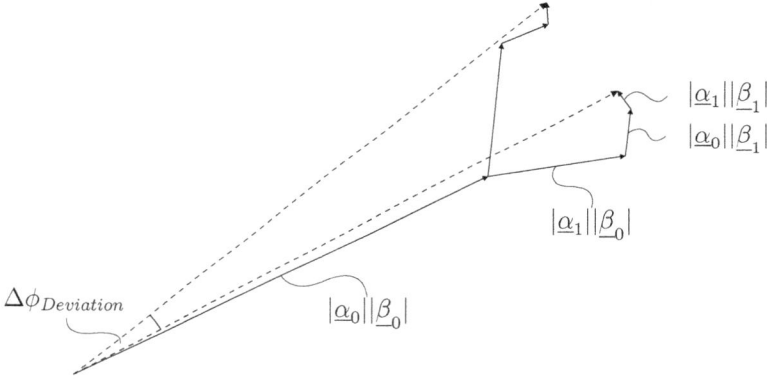

$$|\alpha_1||\underline{\beta}_1|$$
$$|\alpha_0||\underline{\beta}_1|$$
$$|\alpha_1||\underline{\beta}_0|$$
$$\Delta\phi_{Deviation}$$
$$|\alpha_0||\underline{\beta}_0|$$

Figure 5.17: Complex numerator and denominator for an arbitrary scenario

The maximum phase deviation is characterized by a maximum angle between the resulting numerator and denominator vectors. Depending on the length of the constituting vectors, which is defined by the cross-energy of the corresponding complex channel amplitudes $|\underline{\alpha}_i||\underline{\beta}_j|$, a bound for the phase deviation may be derived.

The decisive factor for the maximum phase deviation is the ratio of the sum of all cross-energies involving multipath components to the cross-energy of the corresponding line-of-sight components given as

$$\text{ratio} = \frac{\sum_{i=0}^{I-1}\sum_{j=0}^{J-1}|\underline{\alpha}_i||\underline{\beta}_j| - |\underline{\alpha}_0||\underline{\beta}_0|}{|\underline{\alpha}_0||\underline{\beta}_0|} \tag{5.85}$$

If the cross-energy of the line-of-sight components exceeds the sum of all remaining cross-energies, i.e. ratio ≤ 1, the maximum phase deviation is obtained for a right angle between the multipath vectors and the resulting numerator or denominator vector. An exemplary scenario is depicted in Fig. 5.18.

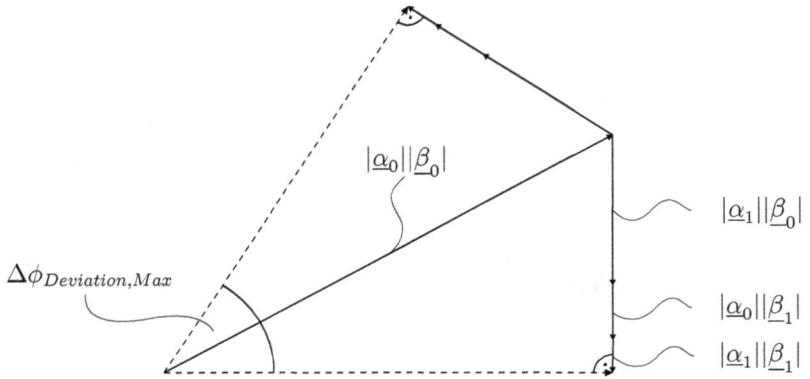

Figure 5.18: Complex numerator and denominator with maximum phase deviation

In this case, the maximum phase deviation can be calculated as

$$\Delta\phi_{Deviation,Max} = 2\arcsin(\text{ratio}) \tag{5.86}$$

The maximum deviation of the time difference of arrival estimate is then given as

$$\Delta\tau_{Deviation,Max} = \frac{\Delta\phi_{Deviation,Max}}{2\pi\Delta f_{Step}} \tag{5.87}$$

Since the numerator and the denominator vector may also occur in exchanged positions, the deviation of the time difference of arrival estimate is bound to the interval

$$-\Delta\tau_{Deviation,Max} \leq \Delta\hat{\tau}_{BA} - \Delta\tau_{BA} \leq \Delta\tau_{Deviation,Max} \tag{5.88}$$

The maximum deviation of the distance estimates for different cross-energy ratios of 1/10 and 1/100 are summarized in Tab. 5.4 and Tab. 5.5, respectively.

Δf_{Step}	200 kHz	400 kHz	600 kHz	800 kHz	1 MHz
$c_0\Delta\tau_{Deviation,Max}$	47.8 m	23.9 m	15.9 m	12.0 m	9.6 m

Table 5.4: Maximum deviation of estimates for a cross-energy ratio of 1/10

Δf_{Step}	200 kHz	400 kHz	600 kHz	800 kHz	1 MHz
$c_0\Delta\tau_{Deviation,Max}$	4.8 m	2.4 m	1.6 m	1.2 m	0.95 m

Table 5.5: Maximum deviation of estimates for a cross-energy ratio of 1/100

If the sum of all multipath cross-energies exceeds the cross-energy of the line-of-sight components, i.e. ratio > 1, the right angle condition is no longer applicable and the phase deviation is unbounded on a 2π-interval. In this scenario, the maximum phase deviation can be defined as

$$\Delta\phi_{Deviation,Max} = 2\pi \qquad (5.89)$$

and no reliable time difference of arrival estimation is possible. An exemplary scenario is depicted in Fig. 5.19.

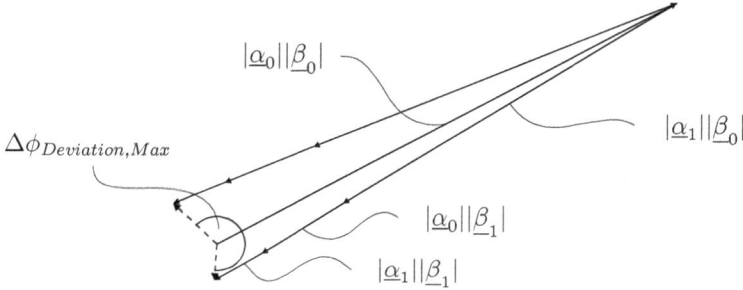

Figure 5.19: Complex numerator and denominator with unbounded phase deviation

5.2.4 Averaging of Burst Phase Differences

The influence of the signal modulation on the phase estimation has been described by the expressions

$$e^{j(\phi(t-\tau_B,\mathbf{b_n})-\phi(t-\tau_A,\mathbf{b_n}))} \quad \text{and} \qquad (5.90)$$

$$e^{j(\phi(t-\tau_{B,j},\mathbf{b_n})-\phi(t-\tau_{A,i},\mathbf{b_n}))} \qquad (5.91)$$

for the simple delay scenario and the multipath scenario, respectively.

These burst phase differences have been replaced by a constant phase term $\Delta\phi_{GMSK}$ or $\Delta\phi_{GMSK,i,j}$ in order to obtain a time-independent term for further evaluation. In the following, the validity of this operation is investigated.

The burst phase differences for five exemplary normal bursts with a relative shift between the signals of $\Delta\tau_{BA}/T_b = 0.1$ are shown in Fig. 5.20. The burst phase differences for one exemplary burst for different values of $\Delta\tau_{BA}/T_b$ are depicted in Fig. 5.21.

For a fixed relative shift $\Delta\tau_{BA}/T_b$, the maximum deviations of the burst phase differences are equal in positive and negative directions. Since all considered burst signals contain the same training or synchronization sequence, the burst phase differences in these parts of the signals are identical. The relative shifts between the signals $\Delta\tau_{BA}/T_b$ have a major influence on the maximum deviation of the burst phase differences from the zero value. This results from the phase modulating nature of the GMSK modulation scheme.

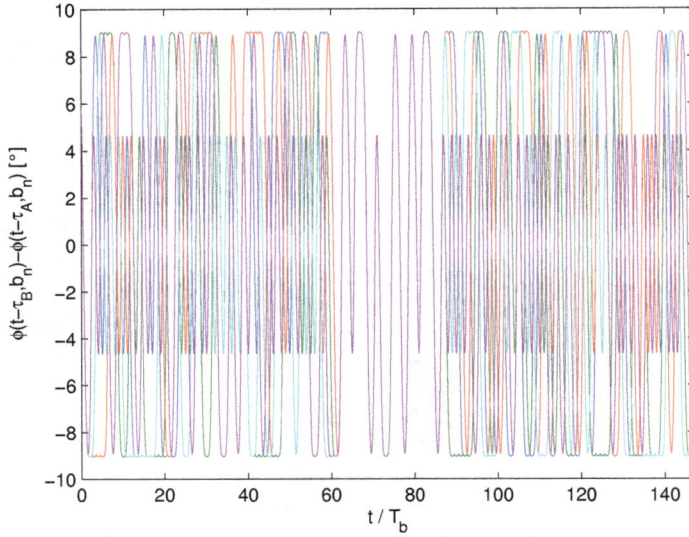

Figure 5.20: Burst phase differences for five normal bursts with $\Delta\tau_{BA}/T_b = 0.1$

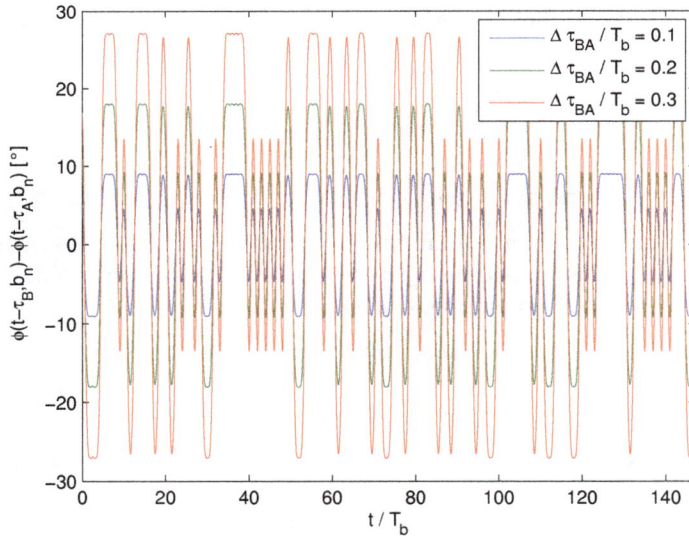

Figure 5.21: Burst phase differences for a normal burst for different values of $\Delta\tau_{BA}/T_b$

For the applicability of the replacement operation, it has to be assured that the mean values are independent of the modulating bit sequence $\mathbf{b_n}$. Furthermore, it is essential that the averaging operation yields a mean value $\Delta\phi_{GMSK}$ or $\Delta\phi_{GMSK,i,j}$ which only depends on the corresponding time difference of arrival $\tau_B - \tau_A$ or $\tau_{B,j} - \tau_{A,i}$.

The independence of the modulating bit sequence $\mathbf{b_n}$ can be assured by limiting the burst signals to the training or synchronization sequence. The mean value of the burst phase differences is therefore independent of the burst index n and may only be influenced by the relative shifts between the signals $\Delta\tau_{BA}/T_b$.

As an example, the burst phase difference for an arbitrary training sequence with a relative shift of $\Delta\tau_{BA}/T_b = 0.1$ is shown in Fig. 5.22.

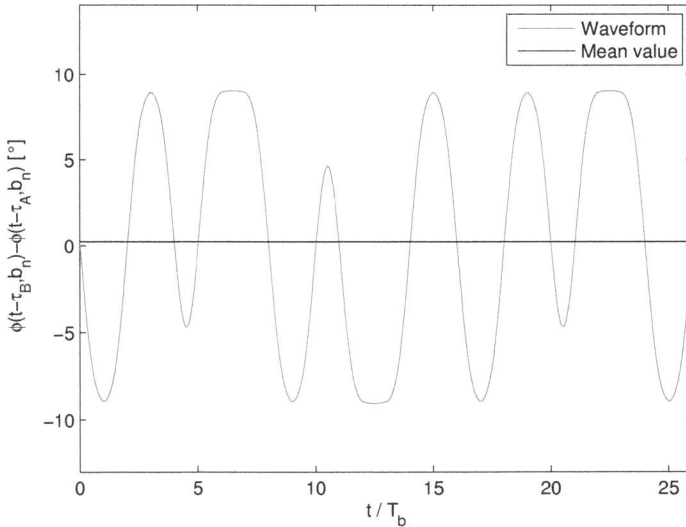

Figure 5.22: Burst phase differences for a training sequence with $\Delta\tau_{BA}/T_b = 0.1$

The mean value of this signal corresponds to $\Delta\phi_{GMSK}$ or $\Delta\phi_{GMSK,i,j}$ according to

$$\Delta\phi_{GMSK} = \frac{1}{T_{Ref}} \int_{t_{Ref,0}}^{t_{Ref,0}+T_{Ref}} \phi(t - \tau_B, \mathbf{b_n}) - \phi(t - \tau_A, \mathbf{b_n}) \, \mathrm{d}t \qquad (5.92)$$

$$\Delta\phi_{GMSK,i,j} = \frac{1}{T_{Ref}} \int_{t_{Ref,0}}^{t_{Ref,0}+T_{Ref}} \phi(t - \tau_{B,j}, \mathbf{b_n}) - \phi(t - \tau_{A,i}, \mathbf{b_n}) \, \mathrm{d}t \qquad (5.93)$$

with T_{Ref} denoting the length and $t_{Ref,0}$ the starting instance of the training or synchronization sequence.

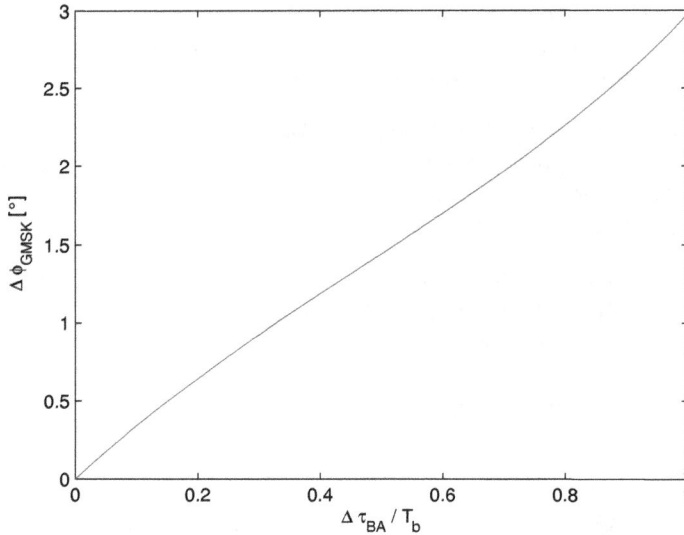

Figure 5.23: Mean value of the burst phase differences for different values of $\Delta\tau_{BA}/T_b$

The dependency of the mean values on the relative shifts $\Delta\tau_{BA}/T_b$ is caused by small averaging imbalances and is depicted in Fig. 5.23 for this exemplary training sequence.

In the case of $\Delta\tau_{BA}/T_b \leq 1$, the mean value is supposed to take small values in the scale of a few degrees and is not expected to have a negative impact on the estimation technique. In the case of $\Delta\tau_{BA}/T_b > 1$, major imbalances caused by the averaging operation have to be expected and no reliable phase evaluation is possible.

5.2.5 Range Ambiguities

The burst phase analysis technique is based on the determination and evaluation the phase differences of burst signals. Since any considered phase difference can only be determined modulo-2π, i.e. the number of full phase cycles of 2π is unknown, the time difference of arrival estimates are only unambiguous within a certain interval.

Considering the simple delay scenario for simplicity, the evaluated quotient for two burst signals is given as

$$\frac{\overline{q}_{n+1}}{\overline{q}_n} = e^{-j2\pi\Delta f_{Step}(\tau_B-\tau_A)} \tag{5.94}$$

The extension for multiple burst signals is based on further quotients with a constant phase decrement of $2\pi\Delta f_{Step}(\tau_B - \tau_A)$.

Due to the phase ambiguities, the same complex exponential numbers and therefore the same time difference of arrival estimates are obtained for

$$-2\pi\Delta f_{Step}(\tau_B - \tau_A) + m2\pi \quad \text{with} \quad m \in \mathbb{Z} \tag{5.95}$$

Therefore, the time difference of arrival estimates

$$\Delta\tau_{BA} + m\frac{1}{\Delta f_{Step}} \quad \text{with} \quad m \in \mathbb{Z} \tag{5.96}$$

are not differentiable by the phase evaluation. The unambiguous ranges $c_0/\Delta f_{Step}$ for different frequency step sizes are summarized in Tab. 5.6.

Δf_{Step}	200 kHz	400 kHz	600 kHz	800 kHz	1 MHz
$c_0/\Delta f_{Step}$	1.5 km	750 m	500 m	375 m	300 m

Table 5.6: Unambiguous range for different frequency step sizes

The phase ambiguities are an inherent property of the burst phase analysis technique. A solution to this challenge may be accomplished by limiting the considered area of operation to smaller sizes than the respective unambiguous range. Consequently, no additional full phase cycles have to be expected. If this condition may not be realized, the number of full phase cycles might by determined by an alternative estimation technique, such as the wideband crosscorrelation technique, beforehand.

5.3 Realizable Frequency Hopping Signal Configurations

For the application of the presented estimation techniques, the frequency hopping capabilities of GSM signals are essential. In this section, the realizable signal configurations for E-GSM 900 systems are investigated.

The E-GSM 900 uplink band comprises an overall bandwidth of 35 MHz and consists of 174 channels of 200 kHz bandwidth each. The respective carrier frequencies are located in the interval [880.2 MHz, 914.8 MHz] with 200 kHz separation.

Depending on the signal generation procedure, i.e. voice and data connection or modified handover procedure, either normal bursts or access bursts are emitted by the mobile station. For both procedures, up to 64 different carrier frequencies may be applied. The maximum number of emitted bursts is unlimited for the voice and data connection. For the modified handover procedure, the maximum number of bursts is limited to 64.

The wideband crosscorrelation technique poses no restrictions on the applied carrier frequencies or the order of the burst signals. A constant frequency step size is advantageous regarding the mathematical derivations. Therefore, a pseudo-random frequency hopping scheme with arbitrary carrier frequencies may be applied. The burst phase analysis technique requires a linearly increasing or decreasing carrier frequency sequence. Consequently, a cyclic frequency hopping scheme with a constant frequency step size is needed in this case.

The frequency step size should be chosen as small as possible in order to circumvent ambiguity and range limitation issues. Furthermore, the overall bandwidth should be chosen as large as possible in order to obtain a high resolution of the estimation techniques. At the same time, the number of bursts largely influences the processing load and should therefore be as small as possible.

In view of these conditions, the maximum performance signal configurations according to Tab. 5.7 are preferred.

Configuration	N	Δf_{Step}	$f_{c,n}$	$B_{Hopping}$	B_{Eff}
64 / 200 kHz	64	200 kHz	891.2 MHz + n200 kHz	12.8 MHz	23.2 MHz
64 / 400 kHz	64	400 kHz	884.8 MHz + n400 kHz	25.4 MHz	46.4 MHz
58 / 600 kHz	58	600 kHz	880.4 MHz + n600 kHz	34.4 MHz	63.1 MHz
44 / 800 kHz	44	800 kHz	880.2 MHz + n800 kHz	34.6 MHz	63.8 MHz
35 / 1 MHz	35	1 MHz	880.4 MHz + n1 MHz	34.2 MHz	63.5 MHz

Table 5.7: Preferred frequency hopping signal configurations

The configurations 64 / 200 kHz and 58 / 600 kHz are perfectly symmetric regarding the center frequency of the E-GSM 900 uplink band of 897.5 MHz. All other configurations are slightly asymmetric due to the realizable channel grid. Shifted configurations are also possible and may be advantageous for reduced attenuation of the radio waves due to lower carrier frequencies. The effective bandwidth is calculated using the approximative formula and assuming a local oscillator frequency which is centered regarding the frequency hopping signal.

Further preferred configurations for GSM 850, DCS 1800 and PCS 1900 systems can be derived accordingly.

6 Numerical Performance Simulation and Evaluation

In the following sections, the performance of the time difference of arrival estimation techniques is investigated by computer simulations.

6.1 Simulation Environment and Parameters

The purpose of the simulation environment is the realistic modeling of the signals and prediction of the estimation technique performance. Therefore, the entire signal flow is implemented in a consistent and repeatable manner.

GSM Signal Generation

The GSM signal generation procedure is based on the mathematical modeling approach described in previous chapters.

The basic GMSK modulation scheme and burst data structure are provided by the GSM-Sim package [58]. The power ramping at the beginning and the end of the burst signals is assumed to be linear. The generation of the frequency hopping signal is accomplished by multiplication of the baseband signal with a complex exponential function.

The *Oversampling Ratio (OSR)* denotes the number of sampling instants within one bit period T_b and is chosen as $OSR = 148$ for all simulations. This corresponds to a sampling rate of $f_s = OSR/T_b \approx 40.1\,\text{MHz}$ which covers the whole uplink bandwidth of the E-GSM 900 band.

In order to evaluate the best attainable performance for operation in the E-GSM 900 band, a frequency hopping signal configuration with 58 access bursts and frequency step size $\Delta f_{Step} = 600\,\text{kHz}$ is applied. This configuration can be characterized by a hopping bandwidth of $B_{Hopping} \approx 34.4\,\text{MHz}$ and an effective bandwidth of $B_{Eff} \approx 63.1\,\text{MHz}$.

The frequency hopping sequence increases linearly in order to enable a direct comparison between the wideband crosscorrelation technique and the burst phase analysis technique.

Radio Channel Signal Transmission

The radio channel is supposed to be linear and time-independent during the duration of the whole frequency hopping signal. Thus, the channel is assumed to be static over the time of $58T_{Frame} \approx 267.7\,\text{ms}$.

The impulse responses $h_{A,ECB}(t)$ and $h_{B,ECB}(t)$ depend on the investigated propagation environment. The received signal is calculated as convolution of the frequency hopping signal with the corresponding impulse response. If necessary, both functions are interpolated by a factor of 10 before convolution and decimated by a factor of 10 after convolution.

Multiple signal sources or interfering signals are not considered in the simulations.

Acquisition of the Signals

The acquisition of the signals is based on a wideband frontend. The local oscillator frequency is supposed to be centered on the frequency hopping signal. The sampling is assumed to be complex with $OSR = 148$ or $f_s = OSR/T_b \approx 40.1\,\text{MHz}$, respectively. The signals are not interpolated.

Antenna effects or frontend distortions are not considered in the simulations.

Pre-Processing of the Signals

The pre-processing of the signals is required for the wideband crosscorrelation technique using stacked and concatenated signals and the application of the burst phase analysis technique.

The basic common step comprises the isolation of the individual burst signals. The isolation in time is based on finding the first sample which exceeds a relative threshold compared to the maximum value. For the simulations, this sample is always the first one. The isolation in frequency is based on the a-priori knowledge of the frequency hopping sequence using a frequency window bandwidth of $B_{Window} = 200\,\text{kHz}$.

The sampling rate of the isolated signals remains unchanged. A decimation after digital down-conversion for baseband signals is not applied.

Implementation of the Wideband Estimation Techniques

The wideband estimation techniques are implemented as described in the corresponding chapter. The sampling rate is the same for all techniques.

The wideband crosscorrelation technique can be applied directly to the received signals without pre-processing. This case is denoted as direct evaluation. Furthermore, the technique may be applied to pre-processed signals in terms of stacking and concatenation. The correlation maximum is interpolated using a parabola fit through the maximum value and two adjacent samples.

The burst phase analysis technique is applied to the isolated burst signals. The evaluation is based on the training or synchronization sequence only. The actual estimation is carried out using a least squares approximation or inverse discrete Fourier transform using zero-padding to 2^{15} samples.

Incoherent Integration Technique

The performance of the presented estimation techniques is compared to a representative state-of-the-art technique in order to provide a comparison between coherent and incoherent estimation approaches. For this purpose, a simplified version of the *Incoherent Integration (ICI)* technique as described in [59, 60] is employed.

The simplified incoherent integration approach is characterized by an incoherent summation over multiple narrowband correlation results according

$$\mathrm{ICI}(\Delta\tau) = \sum_{n=0}^{N-1} |\mathrm{CCF}_n(\Delta\tau)|^2 \tag{6.1}$$

and

$$\mathrm{CCF}_n(\Delta\tau) = \int_{-\infty}^{\infty} r^*_{A,Burst,ECB,n}(t) r_{B,Burst,ECB,n}(t+\Delta\tau)\,\mathrm{d}t \tag{6.2}$$

with $r_{A,Burst,ECB,n}(t)$ and $r_{B,Burst,ECB,n}(t)$ denoting the received narrowband burst signals in receiving station A and B.

The received signals are digitally down-converted to baseband with no decimation, i.e. the sampling rate is identical for all techniques. The correlation maximum is interpolated using a parabola fit through the maximum value and two adjacent samples.

The incoherent integration approach provides the best attainable accuracy of all state-of-the-art techniques and can therefore be regarded as a direct competitor to the presented coherent estimation techniques. Furthermore, the estimation results are reproducible and require the same amount of burst signals.

6.2 Noise Performance in Multipath-Free Scenarios

The estimation performance in noisy and multipath-free scenarios is interesting for evaluating the precision of the estimation techniques. The precision refers to the reproducibility of measurements, i.e. the degree to which repeated measurements show the same results. The performance may be characterized by the distribution of estimates and statistical values such as the standard deviation.

The noisy and multipath-free scenario is employed to determine the standard deviation of the estimates for typical signal to noise ratios of the received signals.

6.2.1 AWGN Channel Model

The noise performance is investigated in noisy environments without multipath propagation. For simplicity, the radio channel impulse responses are modeled as single delta functions with zero delay, i.e.

$$h_{A,ECB}(t) = \delta(t) \quad \text{and} \quad h_{B,ECB}(t) = \delta(t) \tag{6.3}$$

Therefore, the received signals can be expressed as

$$r_{A,ECB}(t) = s_{ECB}(t) + n_{A,ECB}(t) \quad \text{and} \tag{6.4}$$
$$r_{B,ECB}(t) = s_{ECB}(t) + n_{B,ECB}(t) \tag{6.5}$$

The noise signals $n_{A,ECB}(t)$ and $n_{B,ECB}(t)$ are assumed to be drawn from a *White Gaussian Noise (WGN)* process. The signal to noise ratios for both signals are assumed to be equal and are expressed regarding the narrowband burst signals.

6.2.2 Cramér-Rao Lower Bound

As reference for the best attainable estimation performance, the *Cramér-Rao Lower Bound (CRLB)* is investigated in the following. It refers to the minimum variance (or standard deviation) of any unbiased estimator and is therefore directly applicable in multipath-free scenarios. In these cases, the bound can be expressed as

$$c_0 \sigma_{CRLB} = \frac{c_0}{B_{Eff}\sqrt{(2E_{Total}/\mathcal{N}_0)}} \tag{6.6}$$

with E_{Total} denoting the total energy of the signal, $\mathcal{N}_0$ the noise spectral density and B_{Eff} the effective bandwidth of the signal. [45, 61]

The relationship between the total energy to noise spectral density ratio and the respective signal to noise ratios can be stated as

$$E_{Total}/\mathcal{N}_0 = SNR_{Wideband}B_{Wideband}T_{Frame}N \quad \text{or} \tag{6.7}$$
$$E_{Total}/\mathcal{N}_0 = SNR_{Narrowband}B_{Narrowband}T_{Burst}N \tag{6.8}$$

The effective bandwidth of the signal B_{Eff} has a major influence on the best attainable estimation performance and mainly depends on the structure of the frequency hopping sequence.

In the following, the estimation performance is compared with respect to this fundamental bound. In the case of multipath scenarios, more complex considerations are required as described e.g. in [62, 63].

6.2.3 Estimation Standard Deviation

In the following, the simulation results for the noise environment are presented. For this purpose, a Monte-Carlo simulation has been carried out. As a good approximation, the distributions can be identified as Gaussian with zero mean. An exemplary histogram is depicted in Fig. 6.1.

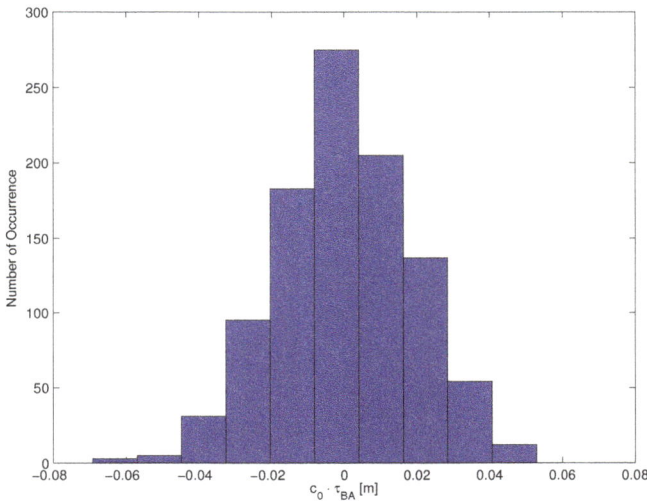

Figure 6.1: Histogram for the wideband crosscorrelation technique with stacked signals at $SNR_{Narrowband} = 10\,\text{dB}$ and 1000 simulation runs

For the further simulations, 100 runs for each signal to noise ratio are carried out. The signal to noise ratio $SNR_{Narrowband}$ refers to the narrowband burst signals. The corresponding standard deviation σ_{TDOA} is used as characteristic value for the distribution.

In real scenarios, the signal to noise ratio may take a variety of values depending on the propagation distance and attenuation of the medium. As a typical value for GSM systems, a signal to noise ratio of $SNR_{Narrowband} = 9$ dB can be assumed which leads to a bit error rate of 10^{-3} for information transmission [64].

Wideband Crosscorrelation Technique

The performance of the wideband crosscorrelation technique is depicted in Fig. 6.2. The case of direct evaluation without pre-processing is compared to the case of pre-processed signals using stacking and concatenation.

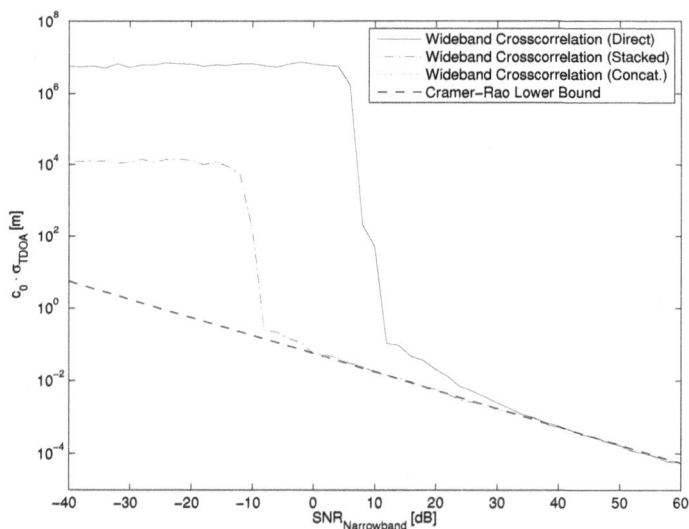

Figure 6.2: Noise performance of the wideband crosscorrelation technique

The pre-processing of the signals results in a considerably improved noise performance compared to the direct evaluation. The pre-processing comprises the isolation of the burst signals and therefore removes the noise from unused time slots and frequency bands. For this reason, the standard deviation is substantially smaller. Furthermore, the wideband crosscorrelation technique using stacking and concatenation perform equally well in noise scenarios.

Regarding the pre-processed signals only, the wideband crosscorrelation technique attains the Cramér-Rao Lower Bound, i.e. yielding the best attainable estimation performance for noise scenarios. For the direct evaluation, this hold true only for high signal to noise ratios. The overall precision for signal to noise ratios $SNR_{Narrowband} \geq 0$ is shown to be below 10 cm.

The wideband crosscorrelation technique using stacked signals is used as representative technique hereafter.

Burst Phase Analysis Technique

The performance of the burst phase analysis technique is depicted in Fig. 6.3. The case of a least squares evaluation is compared to the case of an inverse discrete Fourier transform evaluation.

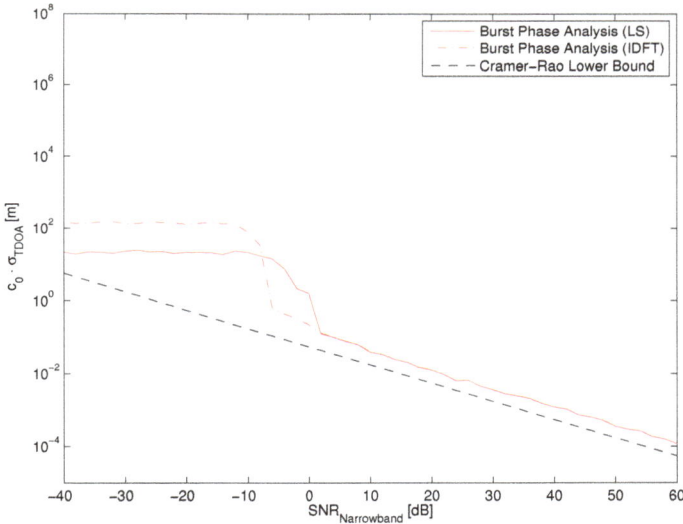

Figure 6.3: Noise performance of the burst phase analysis technique

Both evaluations perform equally well for signal to noise ratios $SNR_{Narrowband} \geq 0$. For reduced signal to noise ratios, the evaluation using the inverse discrete Fourier transform shows gradually better results than the evaluation using the least squares approximation. The technique does not attain the Cramér-Rao Lower Bound. The overall precision for signal to noise ratios $SNR_{Narrowband} \geq 0$ is shown to be below 1 m.

The burst phase analysis technique using an inverse discrete Fourier transform evaluation is used as representative technique hereafter.

Incoherent Integration Technique

The performance of the incoherent integration technique is depicted in Fig. 6.4.

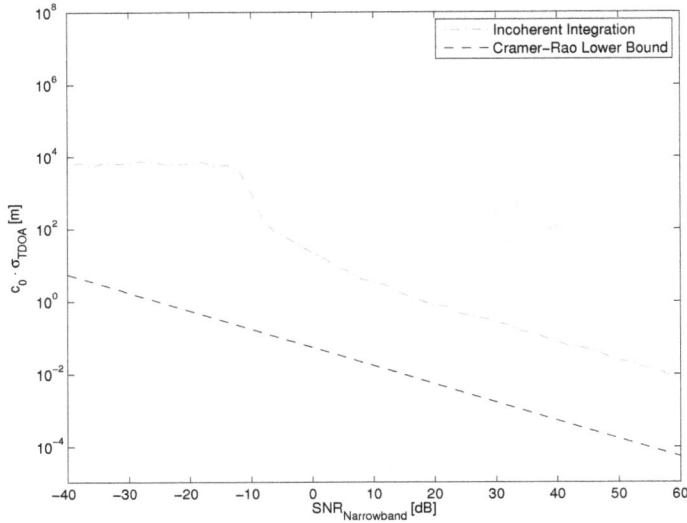

Figure 6.4: Noise performance of the incoherent integration technique

Although the incoherent integration technique is essentially based on crosscorrelation operations, it is not capable of attaining the Cramér-Rao Lower Bound of the wideband signal. This is due to the loss of phase information during the incoherent integration procedure. The overall precision for signal to noise ratios $SNR_{Narrowband} \geq 0$ is shown to be below $100\,\mathrm{m}$.

The loss of phase information is characteristic for all state-of-the-art techniques. The incoherent integration technique therefore enables a representative comparison to the coherent estimation techniques presented in this work.

Comparison of Different Techniques

The performance of the wideband crosscorrelation technique with stacking, the burst phase analysis technique with inverse discrete Fourier transform and the incoherent integration approach is compared in this section. Since pre-processing is inherently included in all these approaches, a fair and balanced comparison is possible. The results are given in Fig. 6.5.

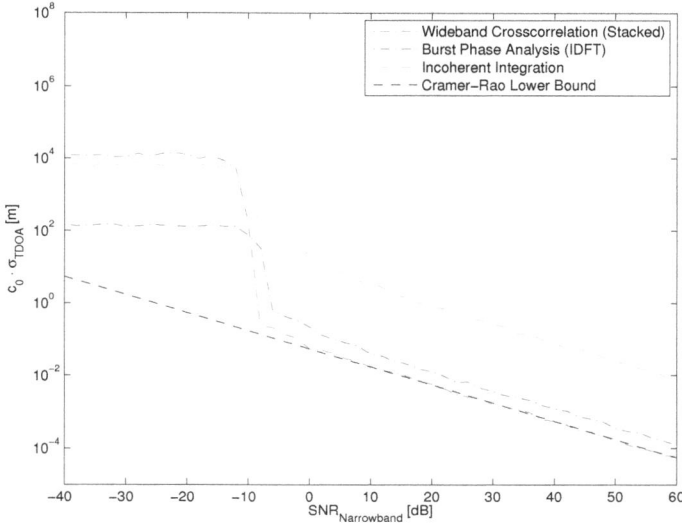

Figure 6.5: Comparison of the noise performance results

For very low signal to noise ratio values $SNR_{Narrowband} \leq -10\,\text{dB}$, the standard deviation of all techniques is very high and is only limited by the discretization of the simulation environment. In this interval, no reliable time difference of arrival estimation is possible.

The interval $-10\,\text{dB} < SNR_{Narrowband} < 0\,\text{dB}$ can be characterized as transition interval where time difference of arrival estimation successively becomes possible. The wideband crosscorrelation technique proves to be the most robust approach in this region.

For signal to noise ratio values $SNR_{Narrowband} \geq 0$, the wideband crosscorrelation technique shows the best performance which is followed by the burst phase analysis technique. The standard deviation of the incoherent integration approach is substantially higher by a factor of approximately 100. From the point of precision in noise scenarios, the wideband crosscorrelation technique with pre-processed signals is therefore the technique of choice.

In general, the noise performance of time difference of arrival estimation techniques can be improved by repeated measurements on the medium and averaging of the obtained results. The mean value successively approaches the desired time difference of arrival value and the overall deviation can be reduced below any required limit.

6.3 Multipath Performance in Two-Paths Scenarios

The estimation performance in two-paths scenarios is interesting for evaluating the resolution and accuracy of the estimation techniques. The resolution refers to the ability of the techniques to differentiate between adjacent propagation paths. It is related to the accuracy, which describes the bias offset from the estimated to the desired value.

The two-paths scenario is employed to determine the maximum estimation bias.

6.3.1 Two-Path Channel Model

The two-paths channel model is characterized by an additional multipath component in one radio channel impulse response. The impulse responses are modeled as

$$h_{A,ECB}(t) = \delta(t) \quad \text{and} \quad h_{B,ECB}(t) = \delta(t) + 0.95\delta(t - \Delta\tau_{MP})e^{j\Delta\phi_{MP}} \quad (6.9)$$

with $\Delta\tau_{MP}$ representing the time difference and $\Delta\phi_{MP}$ the phase difference between the direct component and the multipath component. Furthermore, it is assumed that the multipath component is slightly attenuated with respect to the direct component.

The corresponding crosscorrelation function between the two radio channel impulse responses can then be expressed as

$$h_{A,ECB}(t) \star h_{B,ECB}(t) = \delta(t) + 0.95\delta(t - \Delta\tau_{MP})e^{j\Delta\phi_{MP}} \quad (6.10)$$

The values of $\Delta\tau_{MP}$ and $\Delta\phi_{MP}$ are assumed to be independent variables which characterize the corresponding scenario. The time difference $\Delta\tau_{MP}$ may also be interpreted as spatial difference $\Delta d_{MP} = c_0\Delta\tau_{MP}$.

6.3.2 Maximum Estimation Bias

In the following, the simulation results for the two-paths scenario are presented. For this purpose, the spatial difference Δd_{MP} and phase difference $\Delta\phi_{MP}$ are varied within a regular grid and the corresponding estimation bias $c_0(\Delta\hat{\tau}_{BA} - \Delta\tau_{BA})$ is calculated. The estimation bias is then color-coded in a contour diagram.

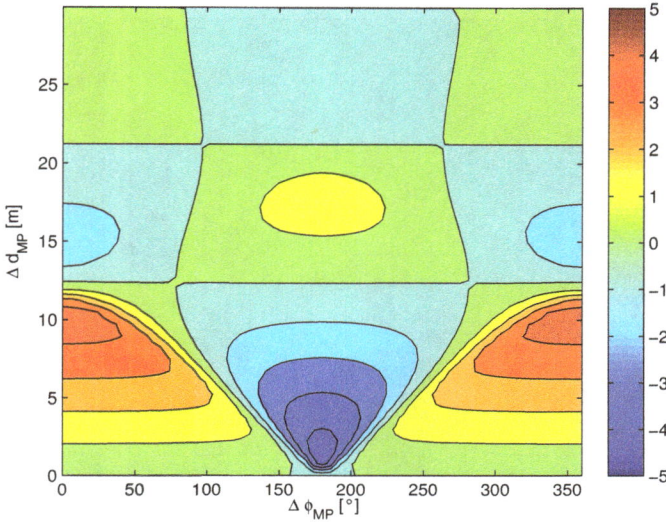

Figure 6.6: Estimation bias of the wideband crosscorrelation technique in meters

Wideband Crosscorrelation Technique

The estimation bias of the wideband crosscorrelation technique is depicted in Fig. 6.6.

The bandwidth of the frequency hopping signal can be approximated as 34.4 MHz yielding a main peak width of approximately 8.7 m. The estimation bias shows to be bound within $-4.5\,\text{m} \le c_0(\Delta\hat{\tau}_{BA} - \Delta\tau_{BA}) \le 4.5\,\text{m}$.

The results are identical for direct evaluation and pre-processed signals using stacking and concatenation. This is due to the identical spectral content of the applied signals which determines the multipath resolvability.

Burst Phase Analysis Technique

The estimation bias of the burst phase analysis technique using the least squares evaluation and the inverse discrete Fourier transform evaluation is shown in Fig. 6.7 and Fig. 6.8, respectively.

The bias for the least squares evaluation shows to be bound within $-5.0\,\text{m} \le c_0(\Delta\hat{\tau}_{BA} - \Delta\tau_{BA}) \le 3.8\,\text{m}$. The bias for the inverse discrete Fourier transform evaluation shows to be bound within $-4.5\,\text{m} \le c_0(\Delta\hat{\tau}_{BA} - \Delta\tau_{BA}) \le 4.5\,\text{m}$.

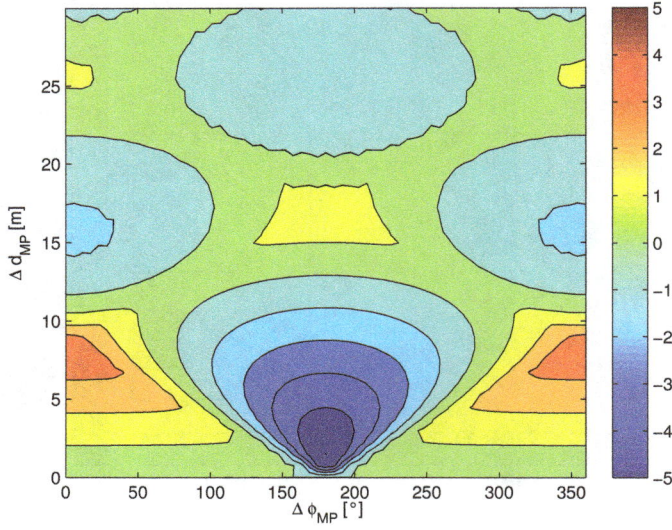

Figure 6.7: Estimation bias of the burst phase analysis technique (LS) in meters

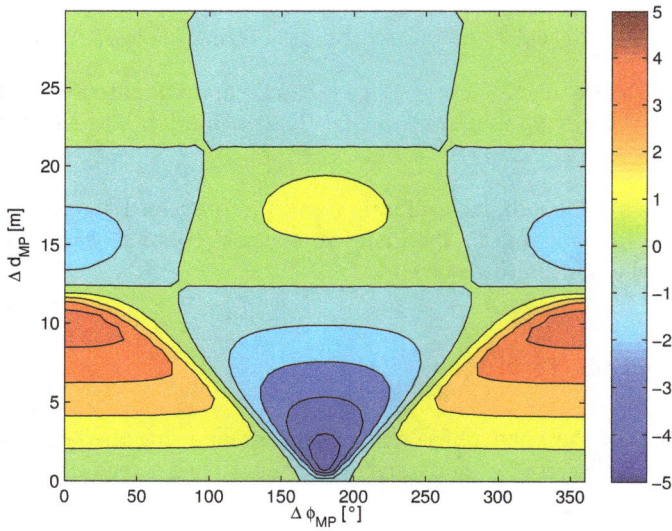

Figure 6.8: Estimation bias of the burst phase analysis technique (IDFT) in meters

As derived for the case of two burst signals, the maximum deviation can be calculated as $2c_0 \arcsin(0.95)/(2\pi \cdot 600\,\text{kHz}) \approx 199\,\text{m}$. This approximation can therefore be regarded as a worst case scenario. The application of multiple burst signals substantially reduces the maximum bias value.

Incoherent Integration Technique

The estimation bias of the incoherent integration technique is depicted in Fig. 6.9.

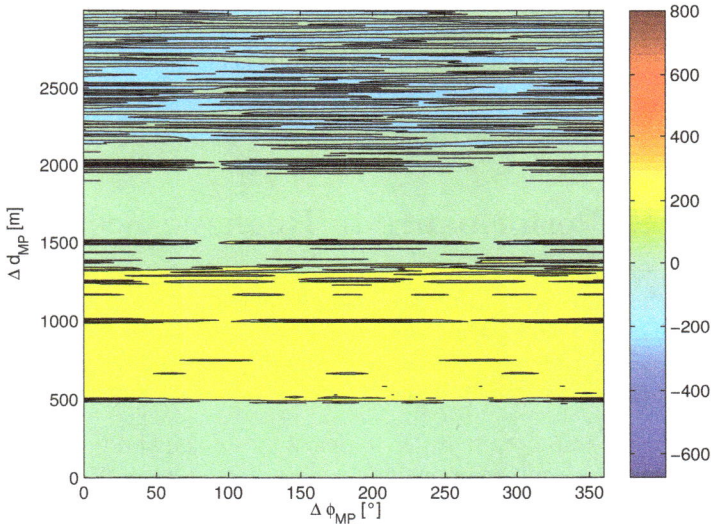

Figure 6.9: Estimation bias of the incoherent integration technique in meters

The estimation bias is characterized by a fundamentally different structure compared to the results presented in previous sections. This is due to the incoherent summation of multiple correlation results. The estimation bias is therefore basically not dependent on the phase difference $\Delta\phi_{MP}$. The scale for spatial difference Δd_{MP} is expanded by a factor of 100.

For an increasing spatial difference Δd_{MP}, the estimation bias increases up to approximately $300\,\text{m}$ and decreases afterwards to approximately $0\,\text{m}$. There are spiky artifacts reaching values between $-673\,\text{m}$ and $853\,\text{m}$.

Comparison of Different Techniques

In the following, the estimation bias of the wideband crosscorrelation technique, the burst phase analysis technique and the incoherent integration technique is compared.

The estimation bias results are similar for all coherent time difference of arrival estimation techniques. The results depend on the spatial difference Δd_{MP} and phase difference $\Delta\phi_{MP}$. The maximum deviation for these techniques is in the scale of $-5.0\,\mathrm{m} \leq c_0(\Delta\hat{\tau}_{BA} - \Delta\tau_{BA}) \leq 5\,\mathrm{m}$.

The estimation bias for the incoherent integration technique shows completely different behavior and is independent of the phase difference $\Delta\phi_{MP}$. The main deviations are in the scale of $0\,\mathrm{m} \leq c_0(\Delta\hat{\tau}_{BA} - \Delta\tau_{BA}) \leq 300\,\mathrm{m}$ with spiky artifacts between $-673\,\mathrm{m}$ and $853\,\mathrm{m}$.

As a main challenge of multipath propagation scenarios, the estimation deviations are not characterized by a zero mean value. Therefore, multipath effects can not be mitigated by averaging of successive estimation results. The estimation bias is a result of the limited resolution of the techniques.

6.4 Multipath Performance in Realistic Scenarios

The estimation performance in realistic scenarios is interesting for evaluating the resolvability of multipath components. The resolvability refers to the ability of the techniques to distinguish between the dominant propagation paths. It is related to the accuracy of the techniques, which describes the bias offset from the estimated to the desired value.

The realistic scenario is employed in order to demonstrate the ability of the techniques to identify individual multipath components.

6.4.1 Tapped Delay-Line Channel Model

For modeling the realistic radio channel impulse responses, a tapped delay-line model is employed. It is characterized by a sum of weighted delta functions according to

$$h_{A,ECB}(t) = \sum_i \alpha_{A,i}\delta(t - \tau_{A,i})\mathrm{e}^{j\phi_{A,i}} \tag{6.11}$$

$$h_{B,ECB}(t) = \sum_j \alpha_{B,j}\delta(t - \tau_{B,j})\mathrm{e}^{j\phi_{B,j}} \tag{6.12}$$

with $\alpha_{A,i}$ and $\alpha_{B,j}$ denoting the amplitudes, $\tau_{A,i}$ and $\tau_{B,j}$ representing the delays and $\phi_{A,i}$ and $\phi_{B,j}$ denoting the phase values. This channel model is suitable for representing static radio channels with a multipath spacing of $1/B_{Signal}$ with B_{Signal} denoting the bandwidth of the applied signal. [34]

In the following, typical *Line-of-Sight (LOS)* scenarios including long and short multipath components are considered.

Line-of-Sight Channel Model with Long Multipath Components

For the line-of-sight channel model with long multipath components, the distance between transmitter and receiver is assumed to be 30 m. The multipath components have long delays with random phase and an amplitude attenuation according to a *Free Space Path Loss (FSPL)* model.

This model describes the attenuation of electromagnetic waves under the assumption of lossless isotropic antennas and can be expressed as

$$\text{FSPL}(d) = \left(\frac{4\pi d f_c}{c_0}\right)^2 \tag{6.13}$$

with d denoting the absolute distance between the transmitter and the receiver and f_c the carrier frequency of 900 MHz. [34]

The exemplary radio channel impulse responses $h_{A,ECB}(t)$ and $h_{B,ECB}(t)$ are depicted in Fig. 6.10 and Fig. 6.11.

Line-of-Sight Channel Model with Short Multipath Components

The line-of-sight channel model with short multipath components is based on the previously introduced channel model. The main difference is the occurrence of multipath components below the resolution limit of the techniques in the scale of 10 m.

The exemplary radio channel impulse responses $h_{A,ECB}(t)$ and $h_{B,ECB}(t)$ are depicted in Fig. 6.12 and Fig. 6.13.

Crosscorrelation of Radio Channel Impulse Responses

The presented techniques are based on the phase coherence between consecutive burst signals and receiving stations. Therefore, only a high accuracy identification of the components of $h_{A,ECB}(t) \star h_{B,ECB}(t)$ is possible. These components are included in the simulation results as desired channel coefficients.

The identification of the radio channel impulse responses $h_{A,ECB}(t)$ and $h_{B,ECB}(t)$ is only possible using narrowband estimation techniques including the training or synchronization sequence. The aforementioned advantages of coherent estimation techniques are not applicable in this case.

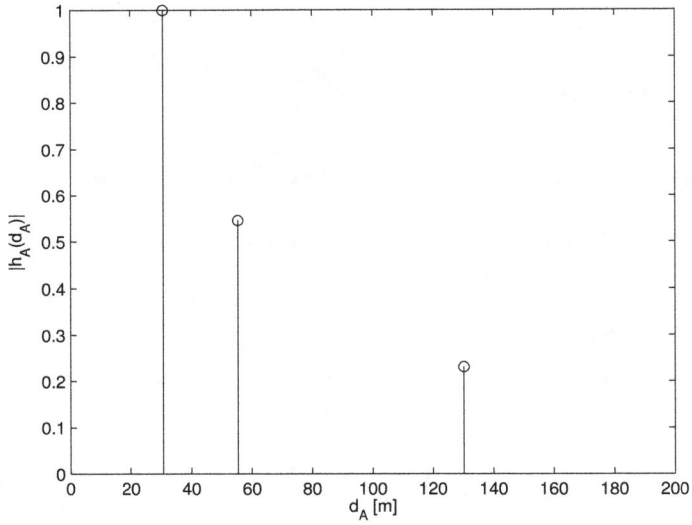

Figure 6.10: Radio channel impulse response $h_{A,ECB}(t)$ for the line-of-sight scenario with long multipath components

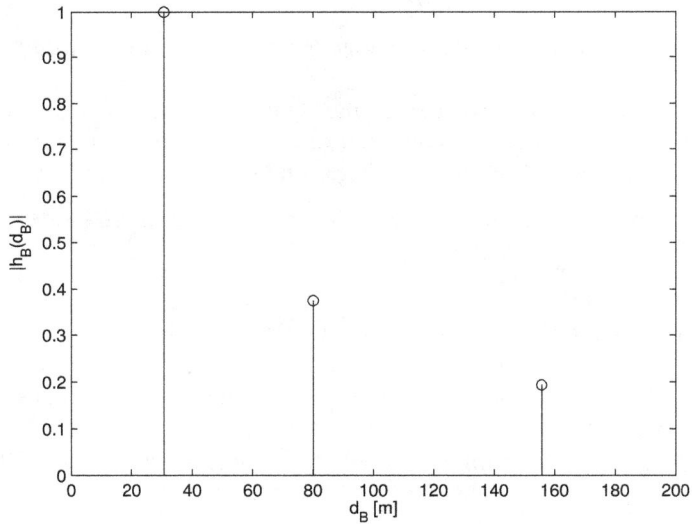

Figure 6.11: Radio channel impulse response $h_{B,ECB}(t)$ for the line-of-sight scenario with long multipath components

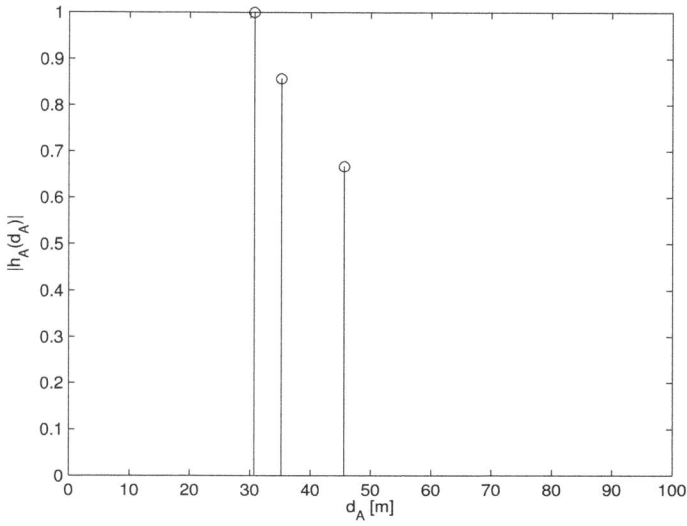

Figure 6.12: Radio channel impulse response $h_{A,ECB}(t)$ for the line-of-sight scenario with short multipath components

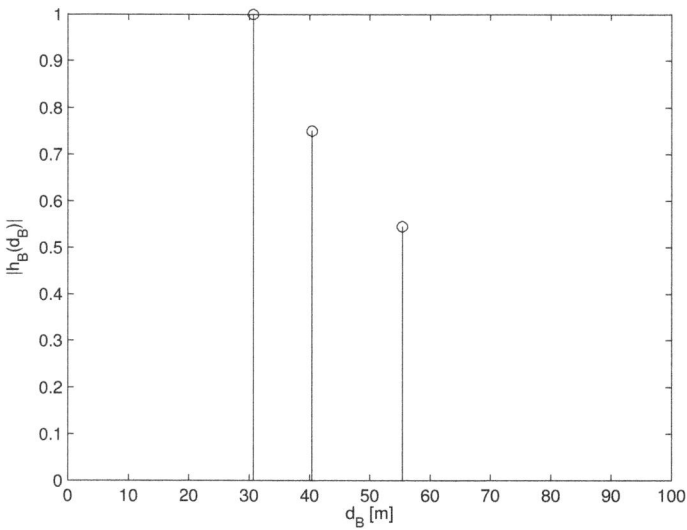

Figure 6.13: Radio channel impulse response $h_{B,ECB}(t)$ for the line-of-sight scenario with short multipath components

6.4.2 Resolvability of Propagation Paths

In the following, the simulation results for the realistic scenarios are presented. For this purpose, the exemplary radio channel impulse responses $h_{A,ECB}(t)$ and $h_{B,ECB}(t)$ are convoluted with the transmitted signal and the techniques for time difference of arrival estimation are applied. Furthermore, the wideband crosscorrelation function $\mathrm{CCF}(\Delta d)$ for the received signals is calculated.

For comparison, the incoherent integration correlation function $\mathrm{ICI}(\Delta d)$ is also included in the figures. Due to the broad correlation peak, this function is only visible for large values of Δd. Since the burst phase analysis techniques are not capable of resolving multipath components, only the resulting estimates are provided.

Line-of-Sight Channel Model with Long Multipath Components

The results for the line-of-sight channel model with long multipath components are given in Fig. 6.14 and Fig. 6.15. In this example, the burst phase analysis technique with least squares evaluation yields an estimate of 5.23 cm. The inverse discrete Fourier transform evaluation results in an estimate of 15.26 cm.

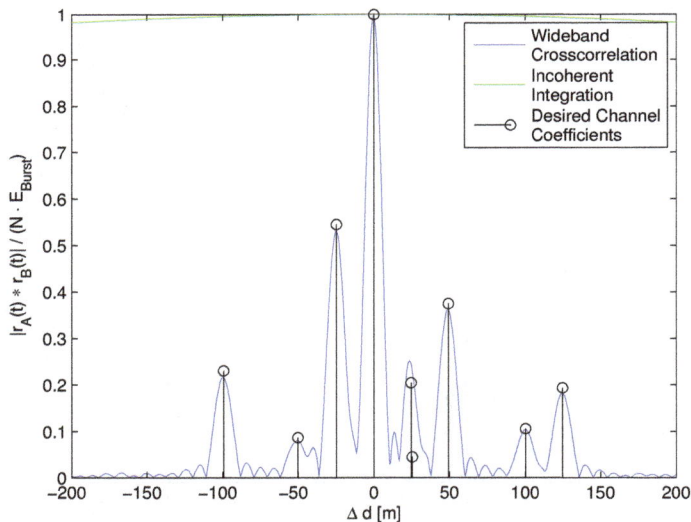

Figure 6.14: Multipath resolvability for the line-of-sight channel model with long multipath components

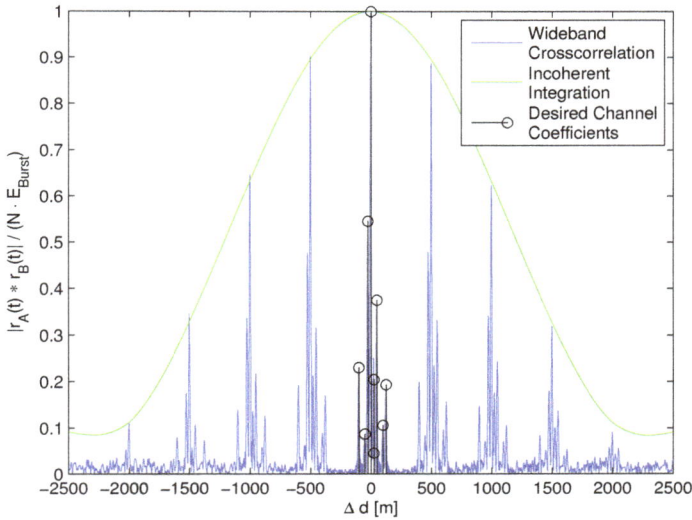

Figure 6.15: Multipath resolvability for the line-of-sight channel model with long multipath components

In this scenario, the wideband crosscorrelation function provides a precise identification of the desired channel coefficients. The peaks of the correlation function are located at the desired channel coefficients. Due to the side peak distance of $c_0/\Delta f_{Step} = 500\,\text{m}$, the identified profile repeats at these positions.

The incoherent integration technique shows a broad correlation peak and is not capable of separating the propagation paths. The incoherent integration correlation function can be interpreted as envelope of the wideband crosscorrelation function for large distances.

The burst phase analysis techniques yield estimation results close to the line-of-sight component $\Delta\tau_{BA}$ which is supposed to equal zero in this scenario. Nevertheless, the individual channel coefficients can not be identified.

Line-of-Sight Channel Model with Short Multipath Components

The results for the line-of-sight channel model with short multipath components are given in Fig. 6.16 and Fig. 6.17. In this example, the burst phase analysis technique with least squares evaluation yields an estimate of $-5.19\,\text{m}$. The inverse discrete Fourier transform evaluation results in an estimate of $2.99\,\text{m}$.

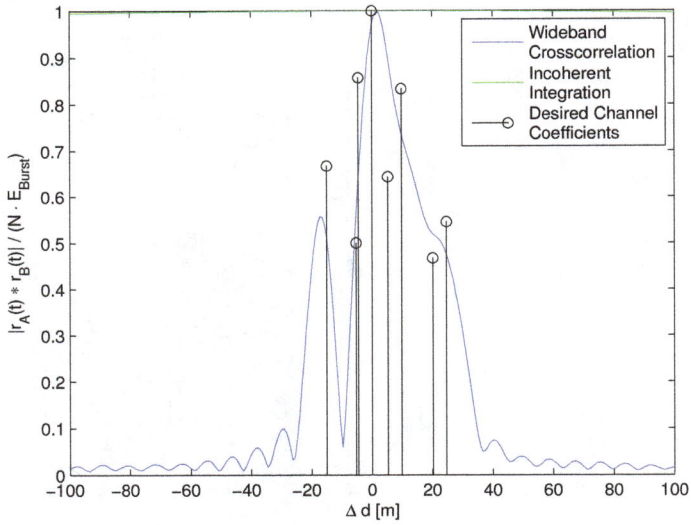

Figure 6.16: Multipath resolvability for the line-of-sight channel model with short multipath components

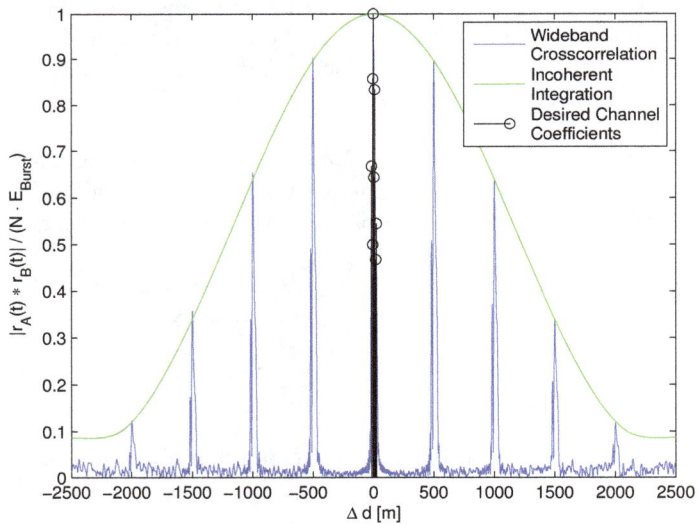

Figure 6.17: Multipath resolvability for the line-of-sight channel model with short multipath components

In this scenario, the multipath components are located below the resolution limit of the techniques of approximately 10 m. Therefore, the multipath separation capabilities of all techniques deteriorate significantly.

The wideband crosscorrelation function peaks interfere and the identification of individual channel coefficients is not possible. However, the main correlation peak is still close to the line-of-sight value of zero.

The incoherent integration technique shows the same characteristics and limitations as described in the previous scenario.

The burst phase analysis techniques yield estimation results with a slight deviation relative to the line-of-sight value and are not capable of resolving the multipath components.

Comparison of Different Techniques

In the following, the multipath resolvability of the wideband crosscorrelation technique, the burst phase analysis technique and the incoherent integration technique is compared.

The wideband crosscorrelation technique enables the identification of the main propagation components due to the narrow main peak of the ambiguity function. The width of the main peak is decisive for the maximum resolvable separation of the propagation paths. In the line-of-sight scenario with long multipath components, all desired channel coefficients are successfully resolved. For short multipath components below the resolvability limit, the coefficients are increasingly covered by superposition and become unresolvable.

The burst phase analysis techniques provide no means for resolving the propagation paths. In the line-of-sight scenario with long multipath components, the overall estimation results are close to the line-of-sight values. In the line-of-sight scenario with short multipath components, deviations in the scale of a few meters have to be expected.

The incoherent integration techniques lacks the ability of resolving multipath components in all investigated scenarios due to the broad correlation peak. However, the maximum of the correlation peak is close to the line-of-sight value.

The performance of all investigated techniques basically depends on the propagation scenario and radio channel impulse responses. All deviations in the results are caused by the radio channel which is unknown a-priori in real applications.

6.5 Summary of Estimation Technique Performance

The coherent time difference of arrival estimation techniques presented in this work
and the incoherent integration technique are compared in Tab. 6.1. The maximum
estimation bias of all coherent time difference of arrival estimation techniques is in the
scale of $\pm 5\,\text{m}$. The incoherent integration technique shows deviations of up to $300\,\text{m}$.

Estimation Technique	Noise Performance	Multipath Resolvability	Computational Load	Reliability
Wideband Crosscorrelation (Direct)	−−	++	−−	++
Wideband Crosscorrelation (Stacked)	++	++	+	++
Wideband Crosscorrelation (Concatenated)	++	++	+	++
Burst Phase Analysis (Least Squares)	0	−−	++	0
Burst Phase Analysis (IDFT)	+	−−	++	0
Incoherent Integration Technique	−	−	0	++

Table 6.1: Comparative summary of coherent and incoherent estimation techniques

The noise performance refers to the standard deviation of the time difference of arrival
estimates. The multipath resolvability denotes the capability of the techniques to identify individual propagation paths. The computational load is estimated based on the
required computing operations, memory requirements and processing time. The reliability refers to the degree of certainty that the obtained result is representative for the
measurement.

In summary, the wideband crosscorrelation technique with stacked signals provides the
best overall performance and can be characterized as technique of choice for applications
in real scenarios.

7 Prototype System Structure and Measurement Results

In the following, the prototype system structure and measurement results are presented. The simulation results are verified and a proof-of-concept is provided.

7.1 Structure of the Receiving Stations

In this section, the main components and the overall structure of the receiving stations is presented. Some major hardware design aspects are based on [65].

7.1.1 Main Functional Components

The receiving stations comprise different components which are essential for the reception of the signals and evaluation of the time difference of arrival between the receiving stations. The main functional components are depicted in Fig. 7.1.

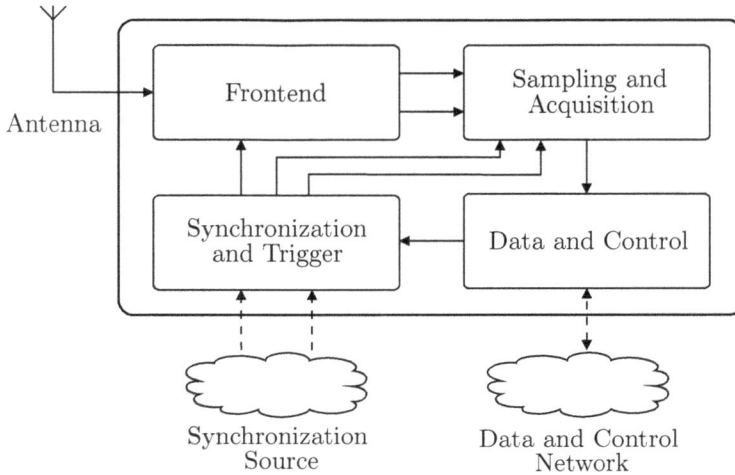

Figure 7.1: Main functional components of a receiving station

The frontend component serves the purpose of receiving the radio frequency signals and is supposed to comprises all required analog components such as amplifiers, filters and mixers. The sampling and acquisition component comprises all required digital components such as analog digital converters and memory to store the acquired signals.

Both components are connected to the synchronization and trigger component, which provides suitable timing and reference signals which are derived from the synchronization signals of the synchronization source. The synchronization of all corresponding local oscillators, analog digital converters and acquisition triggers in all receiving stations is an essential prerequisite for time difference of arrival estimation. The data and control component enables the exchange of system parameters and acquired signals as well as the control of the receiving station components through the data and control network.

The main components and their implementations are described in the following.

7.1.2 Synchronization Source Module

The synchronization source module provides the high frequency and low frequency synchronization signals for the synchronization and trigger module. This functionality is provided using a modified version of the *Local Positioning Radar (LPR)* from Symeo GmbH, Germany [66]. A picture of the module is shown in Fig. 7.2.

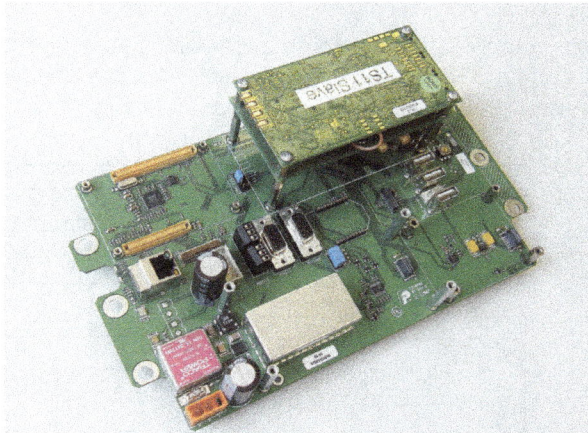

Figure 7.2: Picture of the synchronization source module

The local positioning radar can be characterized as a secondary radar system based on *Frequency Modulated Continuous Wave (FMCW)* signals. It employs frequency chirps of 150 MHz bandwidth in the 5.8 GHz frequency band. The accurate distance measurement between two radar transponders is based on the *Roundtrip Time-of-Flight (RTOF)* principle which establishes a short-time synchronization between the transponders.

By continuous measurements, a long-time synchronization is accomplished. The high frequency synchronization signal of 10 MHz and the low frequency synchronization signal of 1 Hz are then derived from an on-board *Direct Digital Synthesizer (DDS)*. The structure of the synchronization link is depicted in Fig. 7.3.

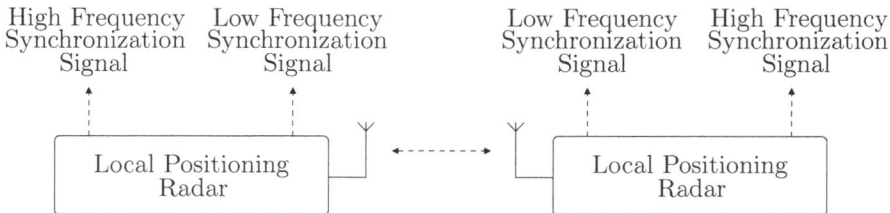

Figure 7.3: Schematic of the synchronization link

For more than two receiving stations, the local positioning radar furthermore enables the establishment of a local coordinate system. Further information on the local positioning radar can be found in [67, 68].

7.1.3 Synchronization and Trigger Module

The synchronization and trigger module serves the purpose of generating synchronous signals for the wideband signal acquisition module. These signals are derived from synchronization signals provided by the synchronization source module. A picture of the custom-designed module is shown in Fig. 7.4.

Figure 7.4: Picture of the synchronization and trigger module

In order to provide synchronization on small time scales and large time scales, the synchronization source provides two synchronization signals. Since frequency and phase synchronization is required, the architecture is based on *Phase Locked Loops (PLLs)*. Furthermore, a logic and control circuitry for trigger release is implemented. The structure of the module is shown in Fig. 7.5.

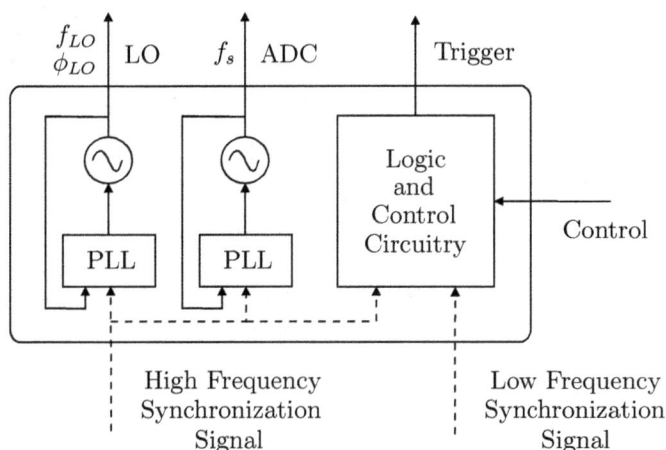

Figure 7.5: Schematic of the synchronization and trigger module

The high frequency synchronization signal is a 10 MHz signal which can directly be converted to the required 900 MHz local oscillator signal and 40 MHz sampling signal. The phase locked loops are based on an integer-N architecture using an adequate divider ratio. The trigger signal is based on a low frequency synchronization signal of 1 Hz or *1 Pulse Per Second (1PPS)*. A logic and control circuitry receives the triggering request within one period of the 1 Hz signal and releases the trigger signal at the next rising edge of this signal. The trigger release procedure is illustrated in Fig. 7.6.

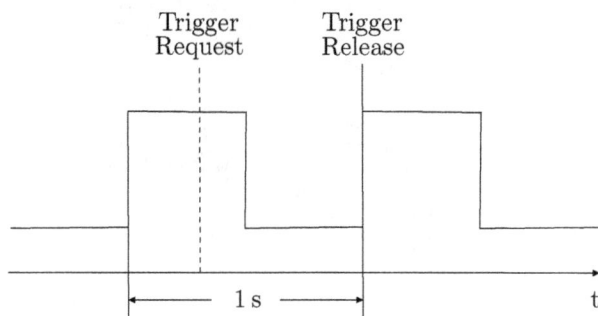

Figure 7.6: Illustration of the trigger release procedure

The synchronization signals with low frequency and high frequency are supposed to be synchronous within all receiving stations. Therefore, the values f_{LO}, ϕ_{LO} and f_s are identical in all receiving stations. Furthermore, the logic and control circuitry ensures a synchronous trigger release in all receiving stations. Further details on the implementation can be found in [69].

7.1.4 Wideband Signal Acquisition Module

The wideband signal acquisition module comprises the frontend component as well as the sampling and acquisition component. This functionality is provided using the ComBlock modules from Mobile Satellite Services Inc., USA [70].

The frontend component is realized using a COM-3007-B wideband receiver with external local oscillator. The sampling and acquisition component is implemented using a COM-8002 high speed data acquisition module with external sampling signal and external acquisition trigger. The interface to the data and control component is realized using a COM-5003 TCP/IP-USB gateway. A picture of the concatenated ComBlock modules is shown in Fig. 7.7.

Figure 7.7: Picture of the wideband signal acquisition module

The frontend component is responsible for the reception of the radio frequency GSM signals. The architecture is characterized as a direct conversion frontend using an IQ mixer. The radio frequency signals are amplified, down-converted and low-pass filtered yielding directly processable complex signals. The acquisition bandwidth of the frontend is determined by the bandwidth of the low-pass filters which are chosen as fourth order elliptic filters with 3 dB cut-off frequency of 20 MHz. A 4 Hz high-pass filter is used to block the direct current bias caused by the local oscillator leakage. The structure is depicted in Fig. 7.8.

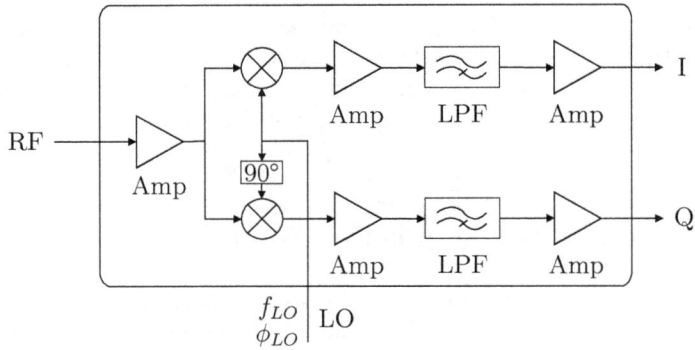

Figure 7.8: Schematic of the frontend module

The full-scale sensitivity for the radio frequency signal is $-51\,\mathrm{dBm}$. The frequency of the local oscillator is chosen near the center frequency of the E-GSM 900 uplink band, in this implementation $f_{LO} = 900\,\mathrm{MHz}$. The phase of the local oscillator ϕ_{LO} is supposed to be identical in all receiving stations.

The sampling and acquisition component is intended for the analog digital conversion and storage of the signals. The analog I and Q signals are sampled by two analog digital converters and are stored in a volatile memory. The structure of the sampling and acquisition component is shown in Fig. 7.9.

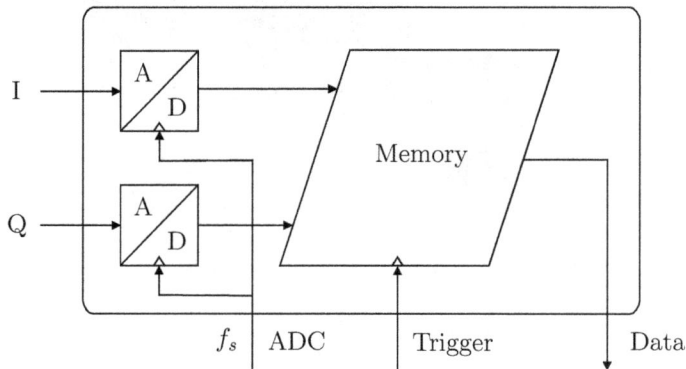

Figure 7.9: Schematic of the sampling and acquisition module

The sampling frequency f_s is supposed to be equal in all receiving stations and is chosen as $f_s = 40\,\mathrm{MHz}$. Furthermore, the rising and falling edges of the sampling signal are synchronized in all receiving stations, i.e. the phase values are equal. The analog digital converters use 10 Bit for quantization. The sampled signals are subsequently stored in a memory of 256 MB capacity whereas the signal acquisition procedure is started by a trigger signal. The triggering edge of the trigger signal is synchronized in all receiving stations. The acquired signals can then be handed over to the data and control module.

7.1.5 Data and Control Module

The data and control module is responsible for the exchange of system parameters and acquired signals as well as the control of the receiving station modules through the data and control network. This functionality is provided using a Pico-ITX computer from Kontron AG, Germany [71]. A picture of the module is shown in Fig. 7.10.

Figure 7.10: Picture of the data and control module

The module is equipped with an Intel Atom Z530 1.6 GHz processor with 2 GB of random access memory. Non-volatile memory is provided using a microSD card. Wired connectivity is enabled using USB and Ethernet interfaces. Wireless connectivity is enabled using an additional wireless local area network adapter. The operating system is based on a Linux system. The structure of the module is depicted in Fig. 7.11.

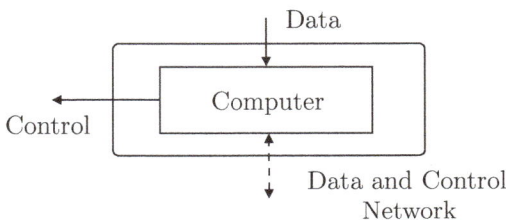

Figure 7.11: Schematic of the data and control module

The computer has access to the stored acquired signals and provides the data to the data and control network. Furthermore, the computer is connected to the synchronization and trigger module for control and initiation of the acquisition trigger.

7.1.6 Integration of the Modules

The main components are integrated into a mobile case for easy handling and deployment. The case includes auxiliary components such as a wideband radio frequency filter, a direct current voltage converter and a lead-acid battery. A picture of the mobile case and the integrated modules is shown in Fig. 7.12.

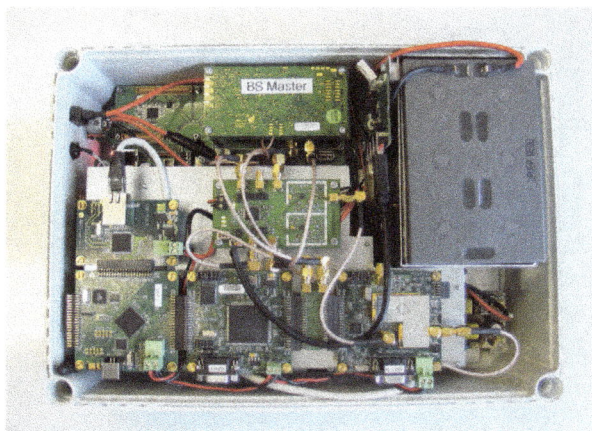

Figure 7.12: Picture of the mobile case and the integrated modules

After starting up, the modules of the receiving station are initialized. As a next step, the receiving station automatically logs on to a personal computer over the wireless local area network connection. The personal computer is responsible for remote control of the receiving stations and the processing of the acquired signals.

A software environment for control of the receiving stations and wireless transfer of the acquired signals to the personal computer has been implemented. The processing of the acquired signals is performed using the MATLAB software package from MathWorks Inc., USA [72]. In particular, the time difference of arrival estimation techniques are evaluated using this software package.

The setup is capable of controlling an arbitrary number of receiving stations. In the scope of this work, two receiving stations have been implemented. Further details on the implementation can be found in [73].

7.2 Characteristics of Hardware Modules

In this section, the characteristics of the hardware modules are presented. The emphasis is on the quality of the synchronization signals and the wideband signal acquisition.

7.2.1 Phase Noise Profile of Synchronization Signals

The phase noise of the synchronization signals is a characteristic profile for each receiving station. For this measurement, a Rohde & Schwarz FSUP signal source analyzer with phase noise measurement option is employed.

The synchronization signals are connected to the $50\,\Omega$ input of the analyzer. The phase noise profile $L(f)$ as well as spurious emissions are evaluated. Furthermore, the corresponding *Root Mean Square (RMS)* timing jitter is calculated. The measurement results for one representative receiving station are presented in the following.

The phase noise profile $L(f)$ and spurious emissions of the local oscillator signal at $900\,$MHz are depicted in Fig. 7.13. The RMS timing jitter is calculated as $4.65\,$ps.

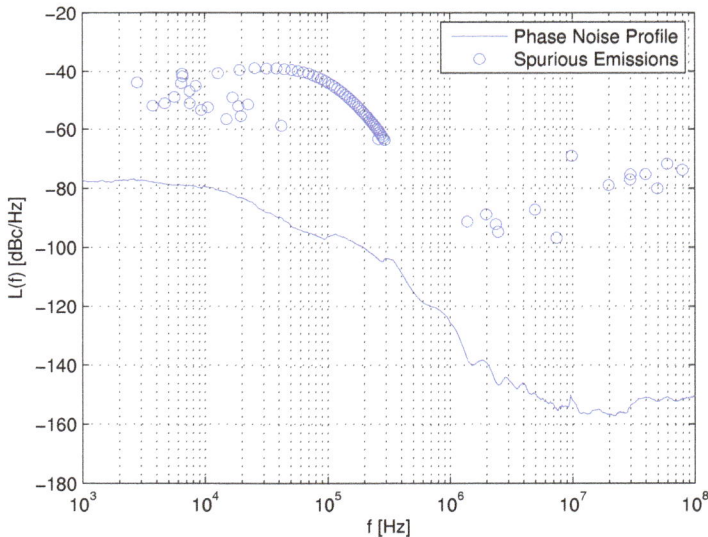

Figure 7.13: Phase noise profile and spurious emissions of the local oscillator signal

The phase noise profile and spurious emissions are noticeable due to the architecture and high amount of digital components of the synchronization and trigger module. The RMS timing jitter is satisfactory and is not supposed to impede the system performance.

The high level of spurious emissions may result in spurious mixing products which may negatively affect the performance of the time difference of arrival estimation.

The phase noise profile $L(f)$ and spurious emissions of the analog digital converter sampling signal at 40 MHz are depicted in Fig. 7.14. The RMS timing jitter is calculated as 8.22 ps.

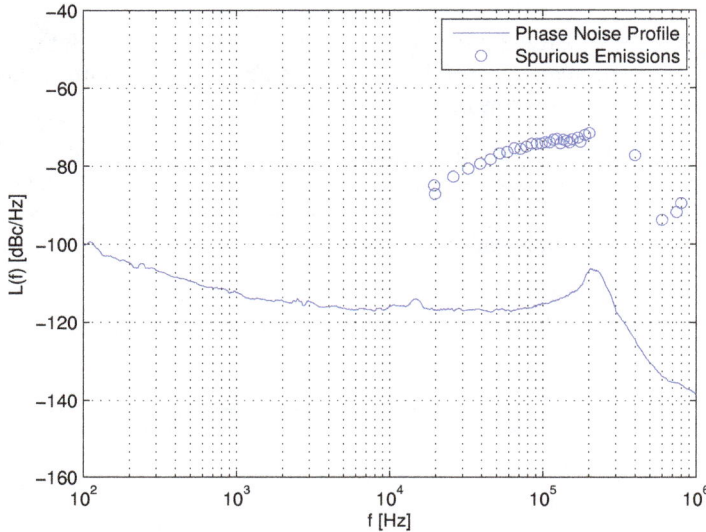

Figure 7.14: Phase noise profile and spurious emissions of the analog digital converter sampling signal

The phase noise profile $L(f)$ and spurious emissions are acceptable for the intended application as sampling signal with sufficiently low levels of phase noise. The high content of spurious emissions is due to the rectangular shape of the sampling signal and is a desired property. The measured values might not be representative since only the fundamental wave has been considered.

The phase noise profile may affect the synchronicity between the receiving stations.

7.2.2 Synchronicity of Synchronization Source Signals

For evaluating the synchronicity of the synchronization source signals, the high frequency and low frequency synchronization signals between two receiving stations are investigated in time domain. For this purpose, a Tektronix TDS6154C digital storage oscilloscope with 15 GHz bandwidth and 40 GS/s sampling rate is employed.

The outputs of the two local positioning radar modules are connected to the $50\,\Omega$ inputs of the oscilloscope for channel A and B, respectively. The synchronization link between the two local positioning radar modules is established using a delay-line cable and a $30\,dB$ attenuator. The trigger of the oscilloscope refers to channel A.

The quality of synchronization is evaluated measuring the timing jitter in terms of timing offset and timing standard deviation between the signals of channel A and channel B. The histograms are characterized as Gaussian distributions.

A snapshot of the high frequency synchronization signal of $10\,MHz$ is shown in Fig. 7.15. The measured timing offset is $0\,s$ and the measured timing standard deviation is $83.3\,ps$. This translates to a phase offset of $0\,°$ and a phase standard deviation of $0.3\,°$.

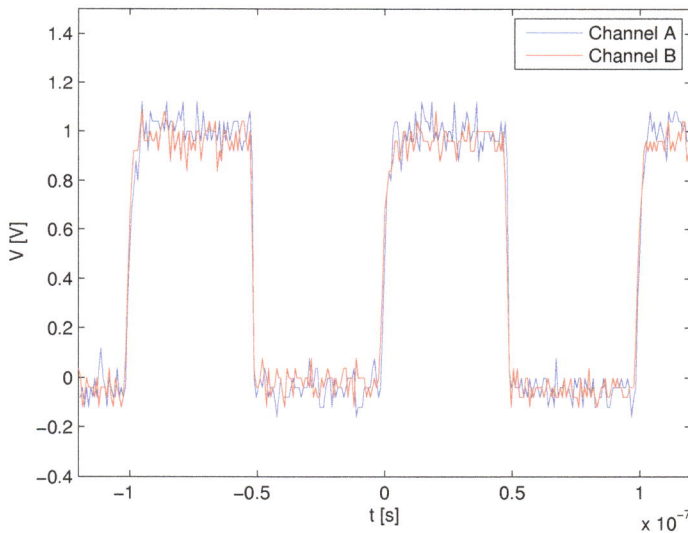

Figure 7.15: Synchronicity of the high frequency synchronization signals

The synchronicity between the high frequency synchronization signals is excellent with zero timing offset and minimal timing standard deviation. Therefore, the local oscillator signal and the analog digital converter sampling signal can be derived from highly synchronous source signals.

A snapshot of the low frequency synchronization signal of $1\,Hz$ is shown in Fig. 7.16. The measured timing offset is $-125\,ps$ and the measured timing standard deviation is $62.5\,ps$. This translates to a phase offset of $-45\cdot10^{-9}\,°$ and a phase standard deviation of $22.5\cdot10^{-9}\,°$.

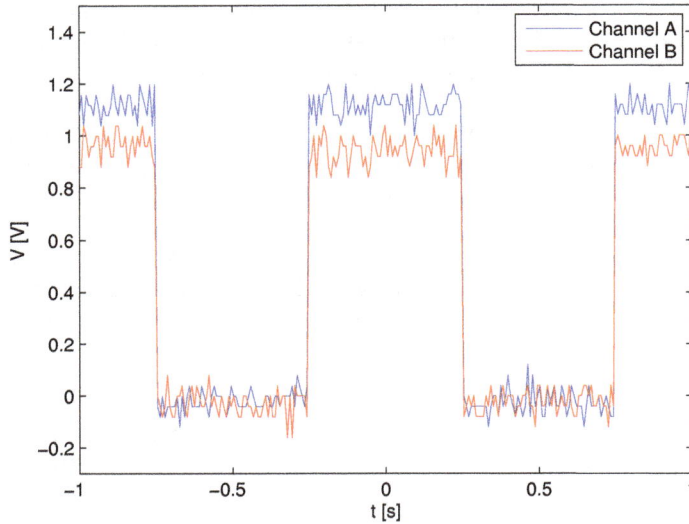

Figure 7.16: Synchronicity of the low frequency synchronization signals

The synchronicity between the low frequency synchronization signals is also excellent with minimal timing offset and minimal timing standard deviation. The derived acquisition trigger signal can therefore be considered as highly synchronous in the receiving stations.

7.2.3 Synchronicity of Synchronization and Trigger Signals

The synchronicity of the local oscillator signals, the analog digital converter sampling signals and the trigger signals between two receiving stations is investigated in time domain using the same measurement setup as previously described.

The synchronization and trigger modules are connected to identical high and low frequency synchronization signals which are provided by two Agilent 33250A arbitrary waveform generators. The output signals are divided by adequate signal splitters.

The quality of synchronization is evaluated measuring the timing jitter in terms of timing offset and timing standard deviation between the signals of channel A and channel B. The histograms are characterized as Gaussian distributions.

A snapshot of the local oscillator signal of 900 MHz is shown in Fig. 7.17. The measured timing offset is 200 ps and the measured timing standard deviation is 43.3 ps. This translates to a phase offset of 64.8 ° and a phase standard deviation of 14.0 °.

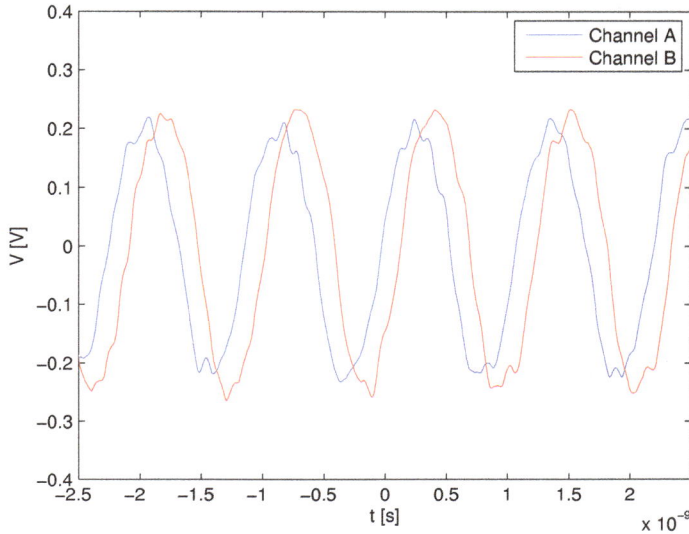

Figure 7.17: Synchronicity of the local oscillator signals

The synchronicity of the local oscillator signals turns out to be a challenging objective. The timing offset is assumed to be caused by small implementation variations between the modules such as connectors, cables or analog and digital components. The timing standard deviation of the signals between the modules may be influenced by the phase noise performance of the individual modules as well as the high level and content of spurious emissions.

A snapshot of the analog digital converter sampling signal of 40 MHz is shown in Fig. 7.18. The measured timing offset is 225 ps and the measured timing standard deviation is 37.5 ps. This translates to a phase offset of 3.24° and a phase standard deviation of 0.54°.

The synchronicity of the analog digital converter sampling signals is satisfactory with acceptable timing offset and timing standard deviation. The rather good performance is supposed to be due to the low divider ratio of the corresponding phase locked loops which has a major impact on the phase noise performance.

Since the trigger signal is released synchronously to the 1 Hz low frequency synchronization signal, the corresponding timing offset and timing standard deviation presented in the previous section are also applicable in this case. The corresponding values have been shown to be minimal and negligible in this implementation.

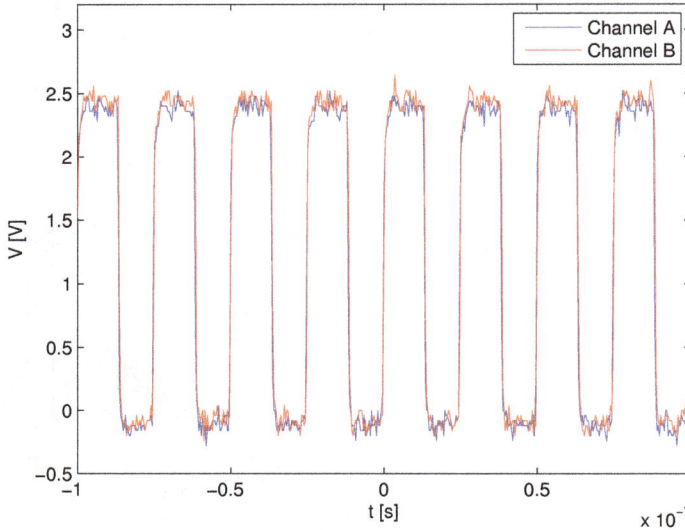

Figure 7.18: Synchronicity of the analog digital converter sampling signals

7.2.4 Frequency Response of Wideband Signal Acquisition Modules

For evaluating the frequency response of the wideband signal acquisition module, a multi-carrier signal with predefined characteristics is injected into the frontend.

The multi-carrier signal comprises 400 sub-carriers using an inter-carrier frequency distance of 100 kHz. The signal covers the frequency band from 880 MHz to 920 MHz with an overall power of −30 dBm. For the generation of this signal, a Rohde & Schwarz SMJ100A vector signal generator is employed.

The local oscillator signal and analog digital converter sampling signal are provided by external sources. For the generation of the local oscillator signal, a Rohde & Schwarz SM300 signal generator is employed. The analog digital converter sampling signal is generated using an Agilent 33250A arbitrary waveform generator.

With this measurement setup, the frequency response $H_{Frontend}(f)$ of the frontend is sampled at a frequency spacing of 100 kHz providing information about the amplitude and phase response at these frequencies. The phase measurements may be inconclusive due to the missing phase reference of the signal source. The amplitude response is depicted in Fig. 7.19 and the phase response is shown in Fig. 7.20.

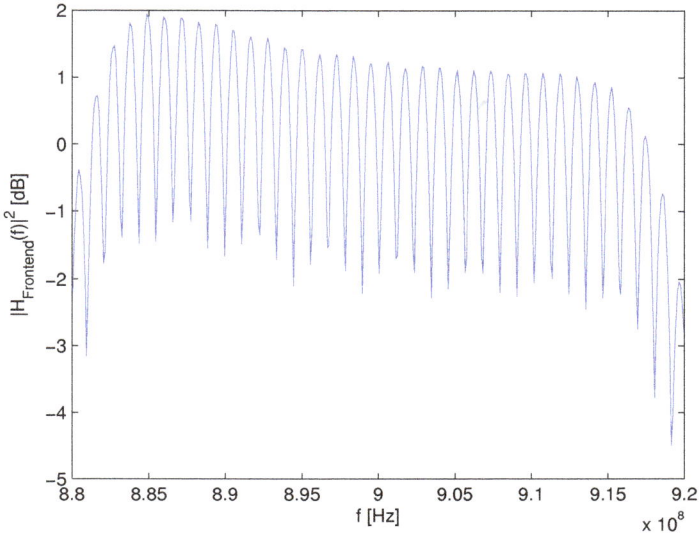

Figure 7.19: Amplitude response of the wideband signal acquisition module

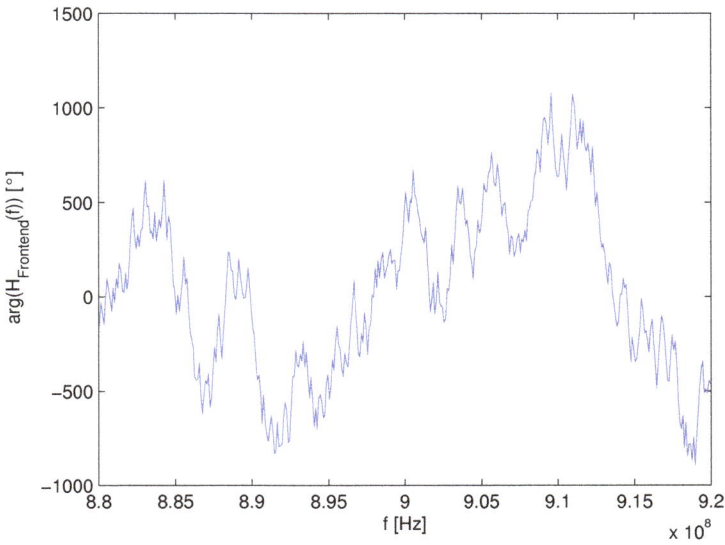

Figure 7.20: Phase response of the wideband signal acquisition module

The amplitude response is characterized by an oscillating nature in the range of $3-4\,\mathrm{dB}$. The mean value is slowly decreasing towards higher frequencies in the scale of $1\,\mathrm{dB}$.

7.2.5 Symmetry of Wideband Signal Acquisition Modules

The symmetry of the wideband signal acquisition modules is characterized by differences in the amplitude and phase response of the frontends.

For evaluating the amplitude and phase response of the frontends, the same multi-carrier signal as described in the previous section is employed. For the generation of this signal, a Rohde & Schwarz SMJ100A vector signal generator is used.

The wideband signal acquisition modules are connected to identical local oscillator signals and analog digital converter sampling signals. The local oscillator signals are provided by a Rohde & Schwarz SM300 signal generator and the analog digital converter sampling signal by an Agilent 33250A arbitrary waveform generator. The output signals are divided by adequate signal splitters.

For a representative comparison between the two wideband signal acquisition modules, the signals are normalized to equal energy. The relative amplitude response of the wideband signal acquisition modules in receiving station A and B is shown in Fig. 7.21.

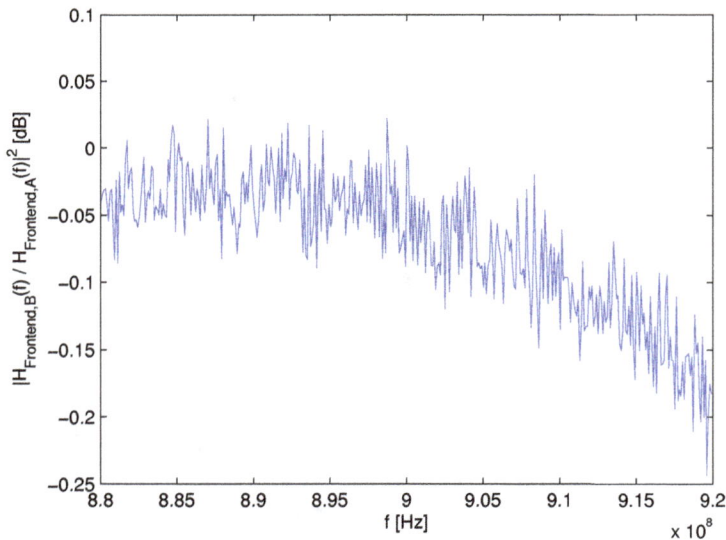

Figure 7.21: Relative amplitude response of the wideband signal acquisition modules in receiving station A and B

The relative amplitude response varies in the scale of $0 - 0.2\,\mathrm{dB}$ with an increasing mean deviation at higher frequencies. This deviation is not supposed to influence the performance of the time difference of arrival estimation techniques.

The relative phase response characterizes the relative phase delay of a carrier signal between the wideband signal acquisition modules. The slope of the relative phase response can be related to the relative group delay between the wideband signal acquisition modules according to

$$\Delta\tau_{Frontend} = -\frac{1}{2\pi}\frac{\mathrm{d}}{\mathrm{d}f}\arg\left(\frac{H_{Frontend,B}(f)}{H_{Frontend,A}(f)}\right) \tag{7.1}$$

The relative phase response of the wideband signal acquisition modules in receiving station A and B is depicted in Fig. 7.22.

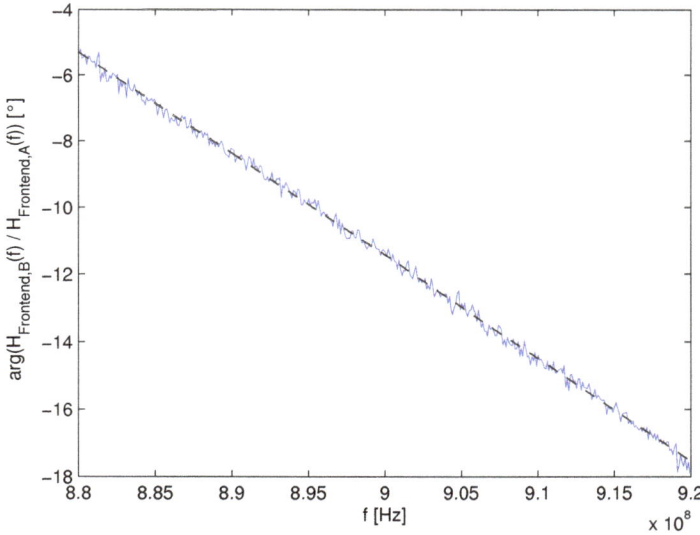

Figure 7.22: Relative phase response of the wideband signal acquisition modules in receiving station A and B

The relative phase response is linearly decreasing with a slope of approximately $-3.05 \cdot 10^{-7}\,°/\mathrm{Hz}$. The linear approximation is included in the figure as a dashed line. The slope corresponds to a relative group delay of $848.2\,\mathrm{ps}$ which is decisive for the reception of modulated signals.

The relative group delay has to be taken into account when evaluating the time difference of arrival between the two receiving stations.

7.3 Time Difference of Arrival Measurement Results

In the following, the static offset compensation procedure as well as the measurement setup and signal configuration for the time difference of arrival estimation are described. Furthermore, the measurement results for the delay-line scenario are presented.

7.3.1 Static Offset Compensation Procedure

Due to the characteristics of the hardware modules, an inherent static offset of the time difference of arrival estimates has to be expected. Based on the measurement results of the individual hardware modules, a compensation of the static offset is possible.

Static estimation offsets are mainly influenced by the synchronization of the local oscillator signals, the analog digital converter sampling signals and the acquisition trigger signals. The influence of synchronization errors has been studied in [74, 75]. Furthermore, the relative group delay of the frontends has to be considered.

Synchronization Source Signals

The synchronicity of the synchronization source signals is excellent with a timing offset of 0 s for the high frequency synchronization signals and −125 ps for the low frequency synchronization signal.

Since the local oscillator signals and analog digital converter signals are derived from the high frequency synchronization signal, no static offsets caused by this signal have to be expected. The acquisition trigger signal is derived from the low frequency synchronization signal resulting in a static offset of −125 ps.

Synchronization and Trigger Signals

The synchronicity of the local oscillator signals and analog digital converter sampling signals is mainly determined by the synchronization and trigger module. The local oscillator signals show a phase offset of 64.8°. The analog digital converter sampling signals show a timing offset of 225 ps.

The phase offset of the local oscillator signals directly translates to a relative phase rotation of the acquired complex signals. This offset is not supposed to impact the time difference of arrival estimation techniques since the techniques are based on the evaluation of phase differences between consecutive burst signals and corresponding receiving stations. A compensation of the phase offset is nevertheless conducted during the evaluation. The timing offset of the analog digital converter sampling signals directly results in a static offset of 225 ps.

Relative Group Delay of the Frontends

The relative group delay characterizes the relative propagation delay of modulated signals between the frontends prior to analog digital conversion. It has been determined as 848.2 ps and has to be considered during the evaluation procedure.

Overall Static Offset

The overall static offset can be determined as sum of all respective offsets as

$$\Delta \tau_{Offset} = \Delta \tau_{Trigger} + \Delta \tau_{LO} + \Delta \tau_{ADC} + \Delta \tau_{Frontend} \quad (7.2)$$
$$= -125 \, \text{ps} + 0 \, \text{s} + 225 \, \text{ps} + 848.2 \, \text{ps}$$
$$= 948.2 \, \text{ps}$$

This value translates to an overall static offset of 0.284 m in the spatial domain.

7.3.2 Measurement Setup and Evaluation Procedure

The measurement setup for verifying the applicability of the presented time difference of arrival estimation techniques is presented in the following. A picture of the complete laboratory measurement setup is shown in Fig. 7.23.

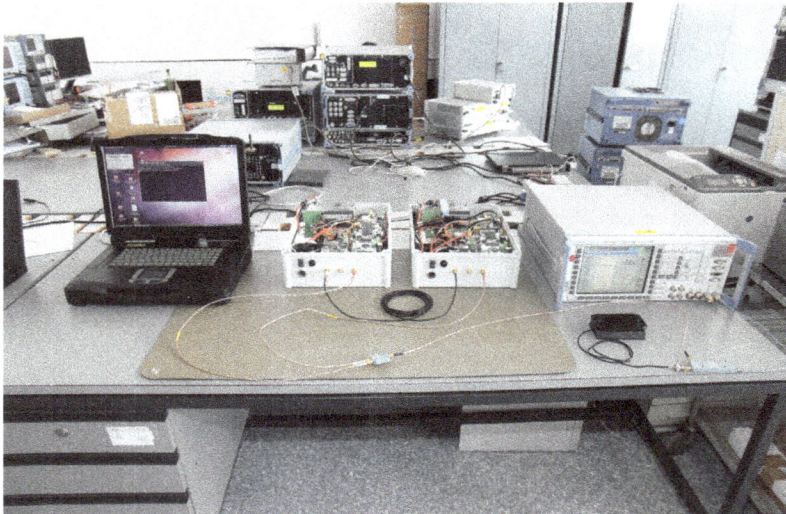

Figure 7.23: Picture of the complete laboratory measurement setup

Frequency Hopping GSM Signal Generation

For generating a realistic frequency hopping GSM signal, a Rohde & Schwarz CMU200 universal radio communication tester in conjunction with a GSM mobile phone is employed.

A voice and data connection using a full-rate traffic channel is established and the uplink signal of the mobile phone is extracted using a directional coupler. A picture of the GSM signal generation setup is shown in Fig. 7.24.

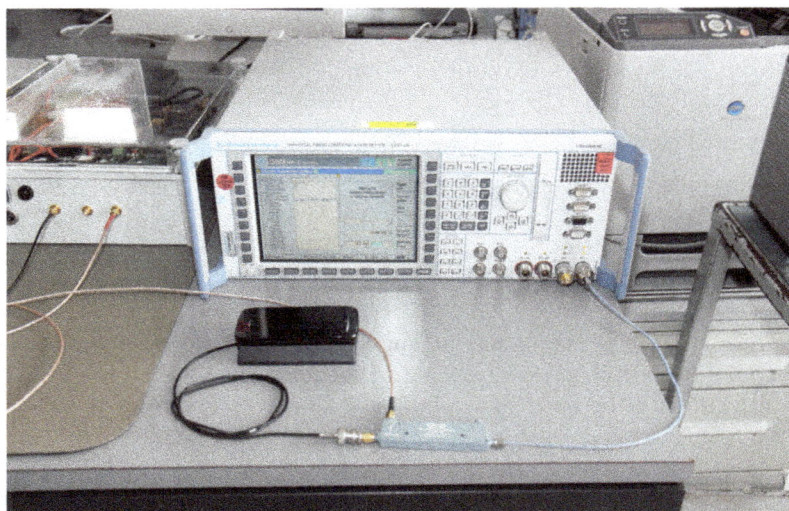

Figure 7.24: Picture of the GSM signal generation setup

Depending on the considered measurement scenario, this signal can be split and combined using coaxial cables in conjunction with power dividers, power combiners and couplers.

Synchronized Wideband Signal Acquisition

The two receiving stations synchronously acquire the corresponding wideband signals which are stored in the on-board memory.

The synchronization link between the two receiving stations is established using a cabled connection. The receiving stations are controlled by a personal computer using a wireless local area network. A picture of the receiving station setup is shown in Fig. 7.25.

Figure 7.25: Picture of the receiving station setup

Evaluation Procedure

The evaluation procedure for determining the time difference of arrival between the receiving station is depicted in Fig. 7.26.

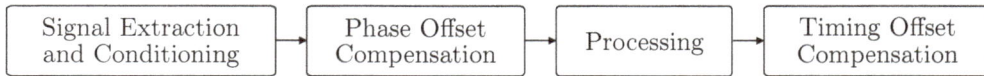

| Signal Extraction and Conditioning | Phase Offset Compensation | Processing | Timing Offset Compensation |

Figure 7.26: Schematic of the measurement evaluation procedure

The signal extraction and conditioning comprises the limitation of the signal duration to the required linearly increasing carrier frequency sequence and the removal of a potential direct current offset of the acquired signals. Moreover, the signals are digitally frequency converted to an ideally centered frequency hopping sequence.

The phase offset compensation accounts for the static phase offsets of the local oscillator signals in both receiving stations. This effect is compensated by multiplication of one acquired signals with a constant phase term of $64.8°$.

The processing of the acquired signals is based on the same parameters as described in the previous chapter. Furthermore, the same algorithms are investigated.

The timing offset compensation is realized by subtracting the determined static offset of $948.2\,$ps from the obtained processing results.

7.3.3 Originating GSM Signal Configuration

The frequency hopping scheme is chosen as cyclic hopping with linearly increasing carrier frequency. The signal comprises 58 normal bursts with a frequency step size of $\Delta f_{Step} = 600\,\text{kHz}$ and a lower starting frequency of $880.4\,\text{MHz}$.

Therefore, this signal corresponds to the simulated signal of the previous chapter except for the burst type. A spectrogram of an extracted and conditioned GSM signal is shown in Fig. 7.27.

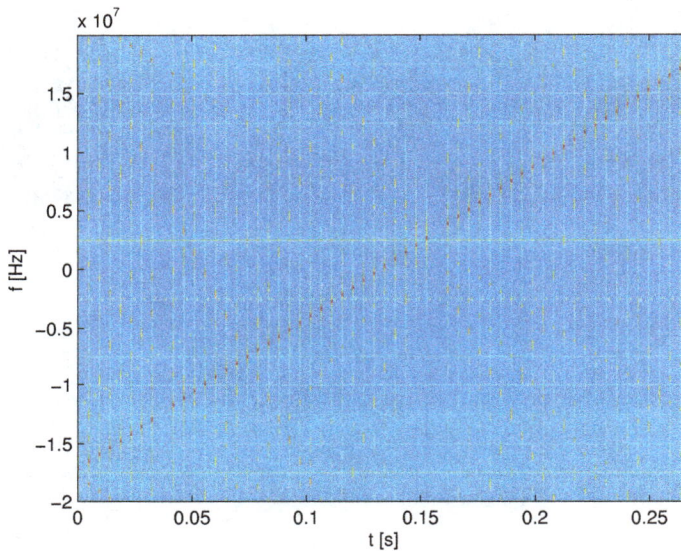

Figure 7.27: Spectrogram of an acquired GSM signal for measurement purposes

This signal serves as originating signal for the following measurements. Due to the full-rate traffic channel configuration, every 26th time frame is idle.

7.3.4 Delay-Line Measurement Results

The delay-line measurement scenario is a multipath-free scenario which is used to verify the general applicability of the time difference of arrival estimation techniques. It is characterized by equal propagation delays of the signal to the receiving stations.

The delay-line cables have an equal length of $1\,\text{m}$. The propagation delay has been verified by a network analyzer and corresponds to $9.6\,\text{ns}$ or $2.9\,\text{m}$ in free space. The measured signal attenuation is $1.2\,\text{dB}$. Due to the equal length and multipath-free propagation, the time difference of arrival $\Delta\tau_{BA}$ is zero. The measurement setup for this scenario is depicted in Fig. 7.28.

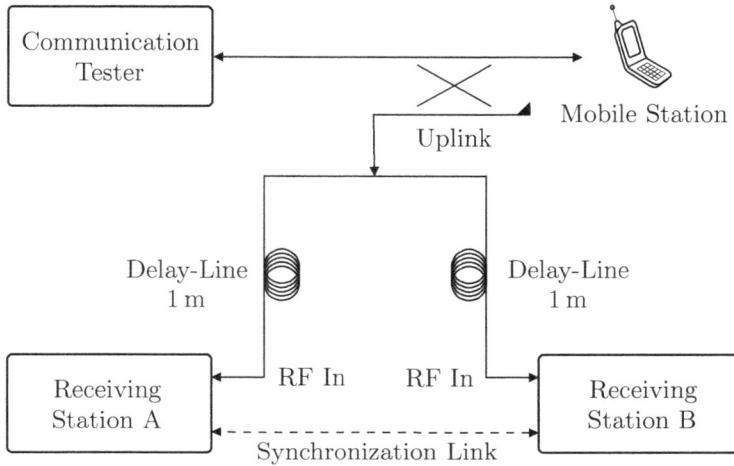

Figure 7.28: Measurement setup for the delay-line scenario

The measurement results for the delay-line scenario are given in Fig. 7.29 and Fig. 7.30. The wideband crosscorrelation technique yields an estimate of 9.2 cm. The burst phase analysis technique with least squares evaluation provides an estimate of 24 cm. The inverse discrete Fourier transform evaluation leads to an estimate of 34 cm. The estimate of the incoherent integration technique shows a large deviation of -7.2 m.[1]

The main peak of the wideband crosscorrelation function is located at the desired channel coefficient attaining the lowest deviation in the centimeter range. The side peak distance of $c_0/\Delta f_{Step} = 500$ m leads to a repetition of the profile at these positions.

Due to unexpected peaks at -23.5 m and 23.7 m, the identification of further potential channel coefficients below these distances is impeded. In particular, potential channel coefficients with attenuation factors below 0.6 can merely be resolved.

The burst phase analysis techniques yield estimation results with slightly higher deviations in the centimeter range. The identification of individual channel coefficients is not possible with these approaches.

The incoherent integration technique exhibits a broad correlation peak. The estimate is severely biased towards the undesired peak of the wideband crosscorrelation function.

For reference, the expected simulation results are given in Fig. 7.31 and Fig. 7.32.

---

[1]This measurement was conducted using an external analog digital converter sampling signal due to start-up issues concerning the wideband signal acquisition module. All other synchronization signals have been provided as described in the previous sections.

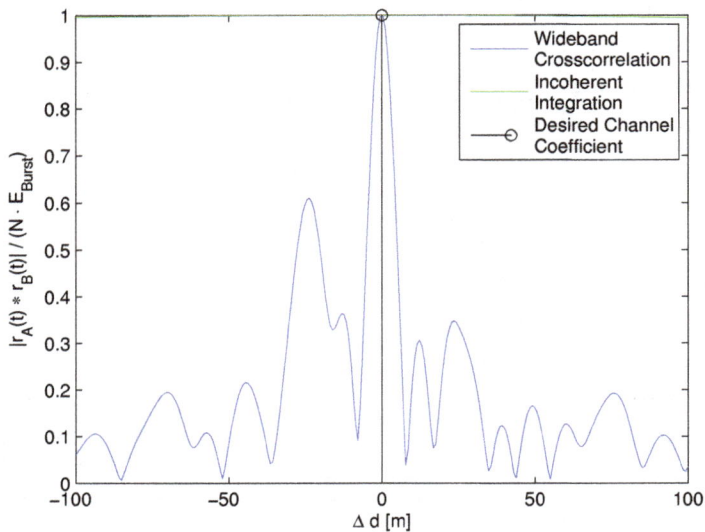

Figure 7.29: Measurement results for the delay-line scenario

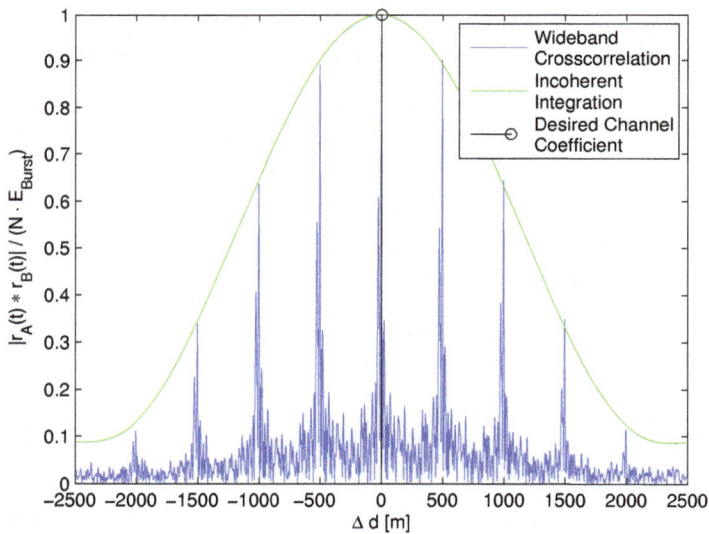

Figure 7.30: Measurement results for the delay-line scenario

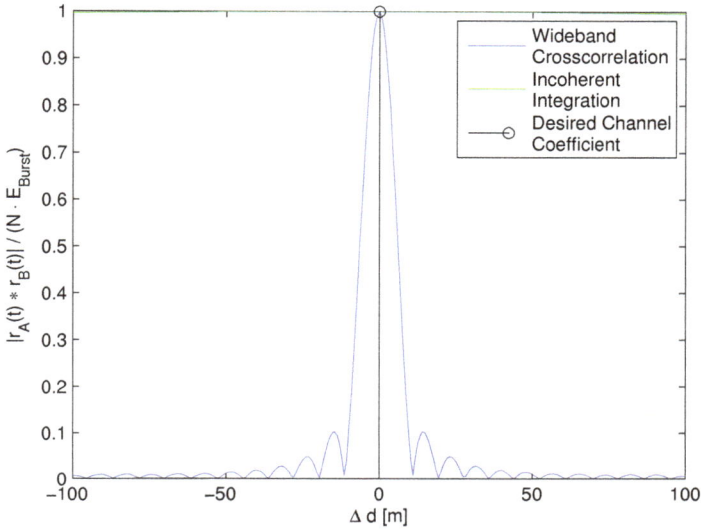

Figure 7.31: Simulation results for the delay-line scenario

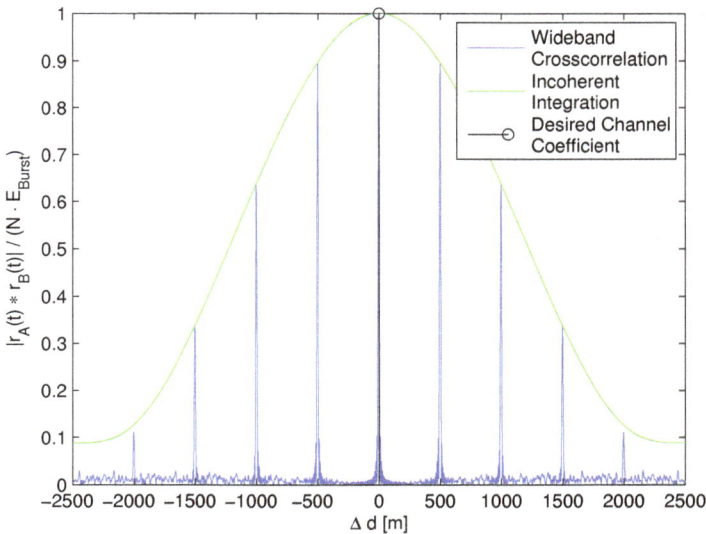

Figure 7.32: Simulation results for the delay-line scenario

7.4 Summary of Hardware and Measurement Results

In the following, the hardware and measurement results are summarized and major conclusions are drawn.

Applicability of Hardware Prototype System

The hardware prototype system comprises the main functional components for time difference of arrival estimation and has been implemented for verifying the applicability of the presented techniques. Therefore, the implementation emphasis has been on providing a proof-of-concept.

The synchronization source module shows excellent performance with minimal offsets and standard deviations of the synchronization signals. These signals can be considered as almost ideal for the desired application and are not supposed to affect the performance of the estimation techniques.

The performance of the synchronization and trigger module is moderate with major issues concerning the local oscillator. The signal is characterized by a high level of phase noise and high content of spurious emissions. Furthermore, the synchronicity between the local oscillators in both receiving stations is challenging. Both effects are due to the structure and implementation of the module and influence the phase coherence of the received signals. The quality and synchronicity of the analog digital converter sampling signal is adequate and is not supposed to influence the performance of the estimation techniques.

The wideband signal acquisition module provides adequate performance for demonstration purposes. The applicability in productive systems might be limited due to its low sensitivity and low dynamic range. Moreover, a limited IQ isolation as well as aliasing artifacts in the presence of strong out-of-band signals have been reported [76].

The data and control module is responsible for the coordination of the functional components and does not affect the quality or synchronicity of the decisive synchronization signals. Therefore, the performance of the estimation techniques is not affected by the data and control module.

A challenging aspect of the hardware prototype system relates to electromagnetic compatibility and electromagnetic interference. Both aspects have been cautiously considered and may influence the quality of the estimation results.

In summary, the presented hardware prototype system is suited for providing a proof-of-concept and verifying the general applicability of the time difference of arrival estimation techniques.

Conclusions from Measurement Results

Based on the measurement results, major conclusions for the prototype system architecture and estimation techniques can be drawn.

Regarding the prototype system architecture, the applicability for time difference of arrival estimation is demonstrated. The described functional components are adequate and the fundamental hardware requirements are accomplished.

The wideband frontend signal acquisition is shown to be feasible for the time difference of arrival estimation techniques. Furthermore, the synchronicity conditions for the local oscillators, analog digital converter sampling signals and acquisition triggers proof to be adequate. The presented trigger release procedure ensures a synchronous signal acquisition in the receiving stations and is feasible for hardware implementation. Moreover, the synchronization source is suitable for the hardware prototype system.

The time difference of arrival estimation techniques show to be applicable for real GSM signals. The fundamental properties of the wideband crosscorrelation function, such as the overall shape and the distances of side peaks, are confirmed. Furthermore, the preprocessing techniques prove to be feasible for the acquired wideband signals. Finally, a rather good correspondence between measurement results, simulations and analytic derivations can be observed.

As expected, the coherent time difference of arrival estimation techniques provide substantially better results than the incoherent estimation technique. Furthermore, the wideband crosscorrelation technique shows better performance than the burst phase analysis techniques. An accurate determination of the time difference of arrival with small deviations in the centimeter range is demonstrated. Furthermore, the existence of a narrow main peak of the wideband crosscorrelation function for multipath resolvability is evident.

Challenging aspects of the obtained measurement results concern the undesired peaks of the wideband crosscorrelation function. These peaks are most likely due to the insufficient quality and synchronicity of the local oscillator signals and prevent a reliable identification of multipath components between $-23.5\,\mathrm{m}$ and $23.7\,\mathrm{m}$ and levels below 0.6 in this case. Repeated measurements show comparable results with different distances of undesired peaks and different peak levels. The estimation results may also be impaired due to electromagnetic compatibility and electromagnetic interference issues.

The evaluation of more sophisticated scenarios with this prototype system in the laboratory is therefore hardly feasible. The simulation results indicate, however, that this challenging aspect is not inherent to the estimation techniques but to the hardware implementation and can therefore be mitigated by a revision of the corresponding functional components.

In summary, a revised prototype system can be assumed to attain the theoretical limits derived in this work achieving an overall accuracy in the scale of $5-10\,\mathrm{m}$.

8 Summary and Outlook

In the following, the major achievements and contributions of this work are summarized. Moreover, the most promising starting-points for future work and development are presented. The work is concluded with an overview over the patents, scientific publications and theses which have been prosecuted, published and supervised during research on this topic.

8.1 Major Achievements and Contributions

The major achievements and contributions of this work can be summarized as follows:

For the first time, the interpretation of a frequency hopping GSM signal as a wideband signal has been demonstrated. Since all state-of-the art techniques for time difference of arrival estimation of GSM signals are essentially based on acquisition and processing of narrowband burst signals, the best attainable accuracy is fundamentally limited. The presented techniques overcome these limitations achieving a localization accuracy in the scale of $5-10\,\mathrm{m}$ as illustrated in Fig. 8.1.

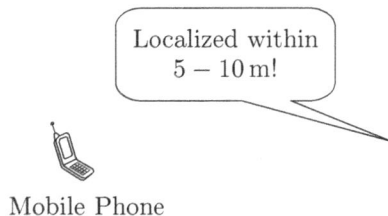

Localized within
$5-10\,\mathrm{m}$!

Mobile Phone

Figure 8.1: Realizable localization accuracy

An essential requirement of the presented techniques is the phase preserving, i.e. coherent acquisition and processing of the signals. For coherent signal acquisition, two possible frontend implementations have been proposed and the corresponding signals have been modeled. Furthermore, two different coherent techniques for time difference of arrival estimation have been analytically derived and investigated.

The performance of the coherent techniques has been compared to a representative incoherent state-of-the-art technique by computer simulations. The main emphasis of the investigations has been on the noise performance and multipath resolvability of the techniques. The performance of the presented coherent techniques has been shown to be superior compared to the incoherent state-of-the-art technique.

The applicability of the presented techniques in real scenarios has been verified by a hardware prototype system. The prototype system architecture has shown to meet the fundamental estimation requirements and the measurement results have provided a proof-of-concept. Further research on this topic is therefore encouraged.

The presented coherent techniques for time difference of arrival estimation permit novel applications with increased accuracy requirements such as highly accurate localization in search and rescue scenarios. Furthermore, the coherent estimation concept can also be adapted to any frequency hopping signal source such as *Terrestrial Trunked Radio (TETRA)*, *Digital Enhanced Cordless Telecommunications (DECT)*, *IEEE 802.15.4 (ZigBee)* and *IEEE 802.15.1 (Bluetooth)* devices.

8.2 Future Work and Development

Future work and development based on this work may focus on the following topics:

Analytical Derivations and Investigations

Both time difference of arrival estimation techniques have been derived for non-moving signal sources. The influence of Doppler shift regarding the ambiguity function of the wideband crosscorrelation technique as well as the quotient series of the burst phase analysis technique may therefore be of interest. Furthermore, the influence of performance degrading mechanisms such as synchronization errors and hardware imperfections may be derived analytically.

For the wideband crosscorrelation technique, the stacking of the received burst signals yields an intermediate multi-carrier signal. The application of state-of-the-art time difference of arrival estimation techniques for multi-carrier signals such as *Orthogonal Frequency Division Multiplexing (OFDM)* signals may therefore be of interest. Furthermore, a reduction of computational load for the crosscorrelation operations may be accomplished by implementation of a zoomed crosscorrelation operation based on a Zoom-FFT or Chirp-Z transform known from spectral analysis.

For the burst phase analysis technique, an analytical derivation using multiple burst signals in multipath scenarios may provide further insight into the resolution of the technique. Furthermore, puncturing methods for idle burst signals due to the 26-multiframe structure of real GSM signals may be investigated.

As further estimation techniques, model-based approaches for channel estimation for frequency hopping communication systems may be of interest. These approaches may be adapted for time difference of arrival estimation purposes. The reliability of these approaches and the necessity of a-priori knowledge of the model order may pose fundamental challenges.

Finally, the influence of coherence bandwidth and coherence time of the radio channel may provide an interesting topic for further investigations. First considerations indicate that a coherent evaluation may only be beneficial for frequency hopping bandwidths below the coherence bandwidth of the radio channel. A partitioning of the frequency bands and incoherent processing afterwards may provide a solution to this challenge.

Improvements and Sophistication of Simulation Environment

The presented performance simulations have been focused on the main aspects of the time difference of arrival estimation techniques such as noise performance and multipath resolvability. The simulation environment may be extended covering further aspects.

The performance of the techniques may be evaluated using sophisticated radio channel models including Doppler shifts and fading effects. Furthermore, typical interference scenarios such as continuous wave interference, blocking scenarios and multiple user interference may be investigated.

The influence of performance degrading mechanisms may also be investigated by simulations. The influence of synchronization errors for the local oscillators, analog digital converters and acquisition triggers may be modeled and implemented in the simulation environment. Furthermore, typical hardware imperfections such as phase noise, spurious emissions, IQ imbalance, quantization, antenna effects and the frontend frequency response may be included in the model.

Hardware Prototype System Enhancements

The prototype system has been implemented in order to provide a proof-of-concept of the presented time difference of arrival estimation techniques. The prototype system may be revised for improved performance.

The quality and synchronicity of the local oscillator signals is a challenging objective. The phase noise profile suffers from the high divider ratio of the employed integer-N phase locked loops. Furthermore, the mixed-signal nature of the synchronization and trigger module causes a high level of spurious emissions. These effects may be mitigated using a different signal generation architecture and additional filtering approaches.

The performance of the wideband signal acquisition module has shown to be moderate. The effects of IQ imbalance may be mitigated during the signal extraction and conditioning procedure. The sensitivity may be improved using additional pre-amplifiers.

The implementation of a narrowband frontend signal acquisition scheme may reduce the required data volume and processing power. However, a coordination and coarse synchronization with the serving base station is necessary in this case.

The extension of signal acquisition for the DCS 1800 or PCS 1900 bands may provide additional frequency hopping bandwidths of 75 MHz or 60 MHz enabling further improvements in localization accuracy.

Additional Measurement Scenarios

In the scope of this work, the basic applicability and performance of the hardware prototype system has been evaluated.

The time difference of arrival estimation performance may be further evaluated in the laboratory using additional networks of delay-lines. The performance in real world scenarios may be predicted using a radio channel emulator in conjunction with sophisticated radio channel models.

Subsequently, the performance of the prototype system may be evaluated in a free space measurement scenario which may require a dedicated GSM base station for initiating the desired GSM signal transmission.

Finally, an entire localization system comprising at least three receiving stations may be implemented and tested in real scenarios.

Further Conceptual Approaches

Using the following conceptual approaches, the entire localization system performance may be improved.

Since a dedicated GSM base station may be required for the application of the presented techniques, the *Timing Advance (TA)* value of the mobile phone may be evaluated for the determination of range ambiguities of the techniques. Thus, the operating range of the system may be extended beyond any required limit.

The presented concept is based on time difference of arrival estimation and provides major enhancements for time-of-flight based localization. The concept may also be adapted for a joint time difference of arrival and angle of arrival estimation using antenna arrays. Furthermore, additional diversity concepts such as polarization diversity may be included in the evaluation. Thus, additional estimates with minimal further hardware may be obtained.

Another topic for further research is the mitigation of estimation deviations in *Non Line-of-Sight (NLOS)* scenarios. These techniques may require an accurate identification of propagation paths and application of further diversity concepts.

8.3 Patents, Scientific Publications and Theses

The following intellectual property rights and patents have been prosecuted during research on this topic:

- A. Götz, Friedrich-Alexander-Universität Erlangen-Nürnberg, "Apparatus and Method for Localization," International PCT Application No. PCT/EP2012/052324, February 10, 2012.

- A. Götz, Friedrich-Alexander-Universität Erlangen-Nürnberg, "Apparatus and Method for Localization," European Patent Application No. EP11154226, February 11, 2011.

- A. Goetz, Friedrich-Alexander-Universitaet Erlangen-Nuernberg, "Apparatus and Method for Localization," US Provisional Application No. US61/441829, February 11, 2011.

The following authored articles have been published and contain excerpts of this work:

- A. Goetz, R. Rose, S. Zorn, G. Fischer, and R. Weigel, "Performance of Coherent Time Delay Estimation Techniques for Frequency Hopping GSM Signals," in *Proceedings IEEE Topical Conference on Wireless Sensors and Sensor Networks*, 2012, pp. 25–28.

- A. Goetz, R. Rose, S. Zorn, G. Fischer, and R. Weigel, "A Burst Phase Analysis Technique for High Precision Time Delay Estimation of Frequency Hopping GSM Signals," in *Proceedings Asia-Pacific Microwave Conference*, 2011, pp. 1446–1449.

- A. Goetz, R. Rose, S. Zorn, G. Fischer, and R. Weigel, "A Wideband Crosscorrelation Technique for High Precision Time Delay Estimation of Frequency Hopping GSM Signals," in *Proceedings European Microwave Conference*, 2011, pp. 33–36.

- A. Goetz, S. Zorn, R. Rose, G. Fischer, and R. Weigel, "A Time Difference of Arrival System Architecture for GSM Mobile Phone Localization in Search and Rescue Scenarios," in *Proceedings Workshop on Positioning, Navigation and Communication*, 2011, pp. 24–27.

The following co-authored articles have been published and cover neighboring topics:

- L. Zimmermann, A. Goetz, G. Fischer, and R. Weigel, "GSM Mobile Phone Localization using Time Difference of Arrival and Angle of Arrival Estimation," in *Proceedings IEEE International Multi-Conference on Systems, Signals and Devices*, 2012.

- S. Zorn, G. Bozsik, R. Rose, A. Goetz, R. Weigel, and A. Koelpin, "A Power Sensor Unit for the Localization of GSM Mobile Phones for Search and Rescue Applications," in *Proceedings IEEE Sensors Conference*, 2011, pp. 1301–1304.

- R. Rose, S. Zorn, A. Goetz, G. Fischer, and R. Weigel, "A New Technique to Improve AoA Using Dual Polarization," in *Proceedings European Microwave Conference*, 2011, pp. 420–423.

- S. Zorn, R. Rose, A. Goetz, R. Weigel, and A. Koelpin, "A New System for Mobile Phone Localization for Search and Rescue Applications," in *Future Security Research Conference*, 2011, ISBN 978-3-8396-0295-9.

- R. Rose, C. Meier, S. Zorn, A. Goetz, and R. Weigel, "A GSM-Network for Mobile Phone Localization in Disaster Scenarios," in *Proceedings German Microwave Conference*, 2011, pp. 1–4.

- S. Zorn, M. Maser, A. Goetz, R. Rose, and R. Weigel, "A Power Saving Jamming System for E-GSM900 and DCS1800 Cellular Phone Networks for Search and Rescue Applications," in *Proceedings IEEE Topical Conference on Wireless Sensors and Sensor Networks*, 2011, pp. 33–36.

- S. Zorn, R. Rose, A. Goetz, and R. Weigel, "A Novel Technique for Mobile Phone Localization for Search and Rescue Applications," in *Proceedings International Conference on Indoor Positioning and Indoor Navigation*, 2010, pp. 1–4.

The following scientific theses have been supervised during the creation of this work:

- M. Arnold, "Verfahren zur Laufzeitdifferenzmessung von schmalbandigen GSM Mobilfunksignalen," Bachelor Thesis, Lehrstuhl für Technische Elektronik, Universität Erlangen-Nürnberg, Germany, 2011, in German Language.

- S. Erhardt, "Implementierung und Evaluierung eines drahtlosen TDOA-Ortungssystems," Bachelor Thesis, Lehrstuhl für Technische Elektronik, Universität Erlangen-Nürnberg, Germany, 2011, in German Language.

- J. Brendel, "Simulation und Entwurf eines Synchronisationsmoduls für ein laufzeitbasiertes Ortungssystem," Intermediate Thesis, Lehrstuhl für Technische Elektronik, Universität Erlangen-Nürnberg, Germany, 2010, in German Language.

- L. Zimmermann, "Laufzeitdifferenz- und Richtungsschätzverfahren zur Ortung von GSM-Mobiltelefonen," Diploma Thesis, Lehrstuhl für Technische Elektronik, Universität Erlangen-Nürnberg, Germany, 2010, in German Language.

- W. Nguatem, "Performance Simulation and Evaluation of TDOA, AOA and Hybrid Localization Schemes," Intermediate Thesis, Lehrstuhl für Technische Elektronik, Universität Erlangen-Nürnberg, Germany, 2010.

Appendices

A Frequency Hopping Sequence Generation in GSM Systems

In the following, the generation algorithm for frequency hopping sequences in GSM systems is described.

The *Mobile Allocation (MA)* table contains all *Absolute Radio Frequency Channel Numbers (ARFCNs)* which are assigned to a mobile station in a GSM cell. Every entry is identified by its *Mobile Allocation Index (MAI)* with MAI = 0 representing the lowest absolute radio frequency channel number in the mobile allocation table.

The algorithm generates an mobile allocation index and thus yields the absolute radio frequency channel number to be used for signal transmission. The main input parameters are:

- *Frame Number (FN)*

- *Hopping Sequence Number (HSN)*

- *Mobile Allocation Index Offset (MAIO)*

N is the number of entries in the mobile allocation table. NBIN is the number of bits required to represent N, i.e. $\text{NBIN} = \text{INTEGER}(\log_2(N) + 1)$.

The variables T1, T2 and T3 are time parameters derived from the frame number. T1R is a reduced version of T1. The relationships are given as follows:

$$T1 = FN \text{ div } (26 \cdot 51) \qquad \text{with } 0 \leq T1 \leq 2047 \text{ (11 bits)} \qquad (A.1)$$
$$T2 = FN \text{ mod } 26 \qquad \text{with } 0 \leq T2 \leq 25 \text{ (5 bits)} \qquad (A.2)$$
$$T3 = FN \text{ mod } 51 \qquad \text{with } 0 \leq T3 \leq 50 \text{ (6 bits)} \qquad (A.3)$$
$$T1R = T1 \text{ mod } 64 \qquad \text{with } 0 \leq T1R \leq 63 \text{ (6 bits)} \qquad (A.4)$$

The *Hopping Sequence Number (HSN)* characterizes the frequency hopping scheme. It is defined as

$$0 \leq \text{HSN} \leq 63 \text{ (6 bits)} \qquad (A.5)$$

with

$$\text{HSN} = \begin{cases} 0 & \text{for cyclic hopping} \\ 1 \ldots 63 & \text{for pseudo} - \text{random hopping} \end{cases} \qquad (A.6)$$

The entry of the mobile allocation table at which the frequency hopping sequence begins is denoted as *Mobile Allocation Index Offset (MAIO)* with

$$0 \leq \text{MAIO} \leq N - 1 \text{ (NBIN bits)} \tag{A.7}$$

The RNTABLE represents a table of 114 integer numbers and is given as follows:

Address	Contents
000...009	48, 98, 63, 1, 36, 95, 78, 102, 94, 73,
010...019	0, 64, 25, 81, 76, 59, 124, 23, 104, 100,
020...029	101, 47, 118, 85, 18, 56, 96, 86, 54, 2,
030...039	80, 34, 127, 13, 6, 89, 57, 103, 12, 74,
040...049	55, 111, 75, 38, 109, 71, 112, 29, 11, 88,
050...059	87, 19, 3, 68, 110, 26, 33, 31, 8, 45,
060...069	82, 58, 40, 107, 32, 5, 106, 92, 62, 67,
070...079	77, 108, 122, 37, 60, 66, 121, 42, 51, 126,
080...089	117, 114, 4, 90, 43, 52, 53, 113, 120, 72,
090...099	16, 49, 7, 79, 119, 61, 22, 84, 9, 97,
100...109	91, 15, 21, 24, 46, 39, 93, 105, 65, 70,
110...113	125, 99, 17, 123

The algorithm is then defined as:

```
if HSN = 0 (cyclic hopping) then:
  MAI, integer (0...N-1) : MAI = (FN + MAIO) modulo N
else:
  M, integer (0...152) :   M  = T2 + RNTABLE((HSN xor T1R) + T3)
  S, integer (0...N-1) :   M' = M modulo (2 ^ NBIN)
                           T' = T3 modulo (2 ^ NBIN)
                           if M' < N then:
                               S = M'
                           else:
                               S = (M' + T') modulo N
  MAI, integer (0...N-1) : MAI = (S + MAIO) modulo N
```

In this algorithm xor represents a bit-wise exclusive or of 8 bit binary operands.

Finally, the absolute radio frequency channel number is calculated according

$$\text{ARFCN} = \text{MA(MAI)} \tag{A.8}$$

A non-hopping frequency hopping sequence is characterized by a mobile allocation consisting of only one radio frequency channel, i.e. $N = 1$ and $\text{MAIO} = 0$. In this case the sequence generation is unaffected by other parameter values.

The stated definitions and more details can be found in [40].

B Logical and Physical Channels in GSM Systems

In the following section, the most important logical channels in GSM and their mapping onto physical channels and resources are described. An overview and further information can be found in [40] and [36].

B.1 Definition of Resources and Channels

For data transmission, the use of physical resources is fundamental. In the case of GSM, a *physical resource* can be characterized by the notation

$$C_n\, T_m \qquad\qquad (B.1)$$

where n denotes the entry index in the *Cell Allocation (CA)* table and m represents the employed *Time slot Number (TN)*. Thus, the applicable resources of a GSM cell are unambiguously defined.

A *physical channel* is an extension of the physical resource concept which also allows a frequency hopping scheme. A physical channel thus defines the physical resources that establish the link between the mobile station and the base transceiver station. In case of a non frequency hopping configuration, the definitions of physical resource and physical channel are equivalent and the notation $C_n\, T_m$ can also be used.

The *logical channel* abstracts from the physical channel and describes the link by means of traffic and control data. The multiplexing of the logical channels, i.e. the traffic channels and control channels, onto the physical channels is organized on the multiframe level and is discussed in the following.

B.2 Structure of the 26-Multiframe

The 26-multiframe consists of 26 subsequent TDMA frames and is intended for user data transmission. It multiplexes the following logical channels onto a physical channel:

TCH The *Traffic Channel (TCH)* carries the user payload. It can be implemented as a full-rate (TCH/F) or half-rate (TCH/H) channel and occupies 24 of the 26 TDMA frames.

SACCH The *Slow Associated Control Channel (SACCH)* is assigned to an established traffic channel and carries slow control data. It uses one of the 26 TDMA frames for the traffic channel full-rate configuration and two for the traffic channel half-rate configuration.

FACCH The *Fast Associated Control Channel (FACCH)* occupies one-half of the bits in eight consecutive traffic channel bursts by setting the stealing flags of the corresponding normal bursts. It is used for control data which has to be transmitted urgently.

The last TDMA frame in the 26-multiframe is idle for the traffic channel full-rate configuration and carries the second half of the slow associated control channel for the traffic channel half-rate configuration.

The structure of the 26-multiframe is depicted in Fig. B.1.

TCH	TCH	TCH	TCH	TCH	TCH	TCH	TCH	TCH	TCH	TCH	TCH	SACCH	TCH	TCH	TCH	TCH	TCH	TCH	TCH	TCH	TCH	TCH	TCH	TCH	IDLE/ SACCH

Figure B.1: Structure of the 26-multiframe

B.3 Structure of the 51-Multiframe

The 51-multiframe is formed of 51 consecutive TDMA frames and is intended for control data transmission which is not directly associated to a traffic channel (i.e. all except the slow associated control channel and the fast associated control channel). It serves the purpose of multiplexing the following logical channels onto a physical channel:

FCCH The *Frequency Correction Channel (FCCH)* is a downlink channel and serves the purpose of frequency synchronization of the mobile station's local oscillator. It is implemented as a sequence of zero bit values which translates to a continuous-wave signal with 67.7 kHz offset from the carrier signal.

SCH The *Synchronization Channel (SCH)* is a downlink channel and broadcasts information to identify a base transceiver station by means of the *Base Station Identity Code (BSIC)*. Furthermore it enables frame synchronization of the mobile station by transmitting the *Reduced Frame Number (RFN)* of the TDMA frame.

BCCH The *Broadcast Control Channel (BCCH)* is a downlink channel and transmits a series of information elements which characterize the organization of the radio network. These elements include information about the radio channel configuration of the current and neighboring cells, synchronization information and registration identifiers. Furthermore it broadcasts information about the structure of the common control channel.

CCCH The *Common Control Channel (CCCH)* comprises the *Paging Channel (PCH)*, the *Access Grant Channel (AGCH)* and the *Notification Channel (NCH)*. The paging channel is used for paging to find a specific mobile station. The access grant channel is employed for assigning a traffic channel or slow dedicated control channel to a mobile station. The notification channel informs mobile stations about incoming group and broadcast calls.

RACH The *Random Access Channel (RACH)* is used in uplink transmission to ask for a dedicated signaling channel for exclusive use by the mobile station.

SDCCH The *Stand-alone Dedicated Control Channel (SDCCH)* is a bidirectional channel which is used for signaling between the mobile station and the base transceiver station when there is no active data connection. It is requested by the mobile station on the random access channel and assigned via the access grant channel.

Since some of the control channels are only used in uplink or downlink transmission, the structure of uplink and downlink mapping is different. For some configurations, two multiframes are needed to map all logical channels. [36]

An important aspect is that the broadcast control channel has always to be transmitted on the first time slot and first frequency assigned to the cell and must not use a frequency hopping scheme. The signal transmission is therefore only allowed on physical channel $C_0 T_0$. The stand-alone dedicated control channels may be assigned to different physical channels.

The random access channel is accessed by the mobile stations in a cell without reservation in a competitive scheme based on the slotted ALOHA principle [36]. It relies on an acknowledgment that the sent signal has been received without collisions. If the reception fails, the access bursts are retransmitted in the next TDMA frame. Since the base transceiver station has not yet had the opportunity to estimate the distance to the mobile station, i.e. the *Timing Advance (TA)* value has not been set yet, the access bursts are not exactly aligned to the TDMA frame structure.

There are many different mapping configurations which can be chosen according to the cell requirements. A minimalist example configuration which can be used in small cells and requires just one physical channel for all required logical channels is the $C_0 T_0$ mapping configuration V [77, 78, 40]. It is a combined configuration in the sense that the stand-alone dedicated control channels are multiplexed with the FCCH, SCH, BCCH, CCCH and RACH on the same physical channel [77].

The structures of the downlink and uplink are depicted in Fig. B.2 and Fig. B.3.

FCCH	SCH	BCCH	BCCH	BCCH	BCCH	CCCH	CCCH	CCCH	CCCH	FCCH	SCH	CCCH	CCCH	CCCH	CCCH	CCCH	CCCH	CCCH	FCCH	SCH	SDCCH(0)	SDCCH(0)	SDCCH(0)	SDCCH(0)

SDCCH(1)	SDCCH(1)	SDCCH(1)	SDCCH(1)	FCCH	SCH	SDCCH(2)	SDCCH(2)	SDCCH(2)	SDCCH(2)	SDCCH(3)	SDCCH(3)	SDCCH(3)	SDCCH(3)	FCCH	SCH	SACCH(0/2)	SACCH(0/2)	SACCH(0/2)	SACCH(0/2)	SACCH(1/3)	SACCH(1/3)	SACCH(1/3)	SACCH(1/3)	IDLE

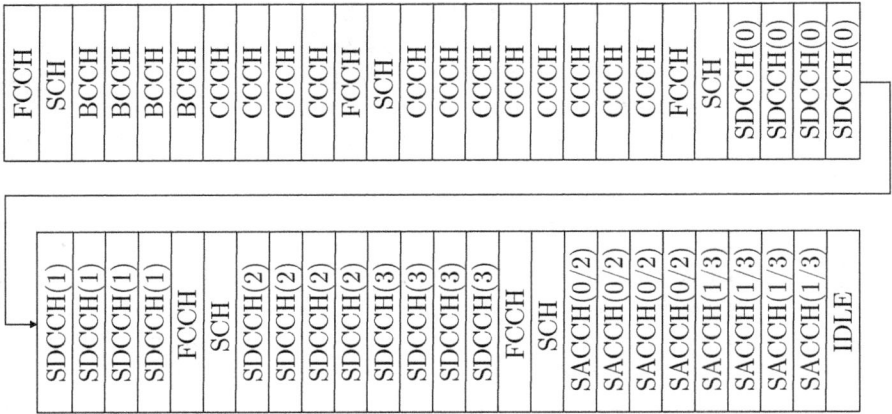

Figure B.2: Downlink structure of $C_0\,T_0$ mapping configuration V

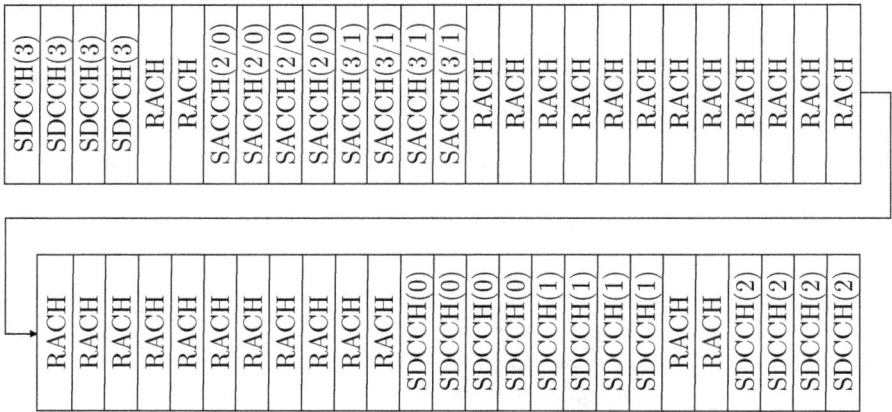

SDCCH(3)	SDCCH(3)	SDCCH(3)	SDCCH(3)	RACH	RACH	SACCH(2/0)	SACCH(2/0)	SACCH(2/0)	SACCH(2/0)	SACCH(3/1)	SACCH(3/1)	SACCH(3/1)	SACCH(3/1)	RACH	RACH	RACH	RACH	RACH	RACH	RACH	RACH	RACH	RACH	RACH

RACH	RACH	RACH	RACH	RACH	RACH	RACH	RACH	RACH	RACH	RACH	SDCCH(0)	SDCCH(0)	SDCCH(0)	SDCCH(0)	SDCCH(1)	SDCCH(1)	SDCCH(1)	SDCCH(1)	RACH	RACH	SDCCH(2)	SDCCH(2)	SDCCH(2)	SDCCH(2)

Figure B.3: Uplink structure of $C_0\,T_0$ mapping configuration V

C Frequency Hopping Signal Transmission Initiation in GSM Systems

For the application of the time difference of arrival estimation techniques, a frequency hopping GSM signal is essential. In the following sections, the two main approaches for initiation of a frequency hopping signal transmission are presented.

C.1 Requirements for Signal Transmission Initiation

According to the GSM standard, there are two basic operational modes for GSM mobile stations which have to be considered:

Idle Mode In idle mode, the mobile station is either turned off, searching for best signal quality broadcast control channels or is already synchronized and ready to perform a random access on the random access channel. [36]

Dedicated Mode In dedicated mode, the mobile station is either occupying a physical channel and trying to synchronize with it or has already established logical channels and is able to switch them through. [36]

For the signal transmission initiation techniques it is assumed that the mobile station is in idle mode and has synchronized to a broadcast control channel. It has already finished the login procedure and is ready to perform a random access on the random access channel.

C.2 Voice and Data Connection

The main idea of this concept is the allocation of a frequency hopping physical channel for voice and data transmission during normal mobile station operation.

Transmission Initiation Procedure

For this purpose, a dedicated bi-directional channel between the *Mobile Station (MS)* and the *Base Station Subsystem (BSS)* has to be established. The necessary procedures are performed by the *Radio Resource (RR)* management. A sequence diagram for this procedure is depicted in Fig. C.1.

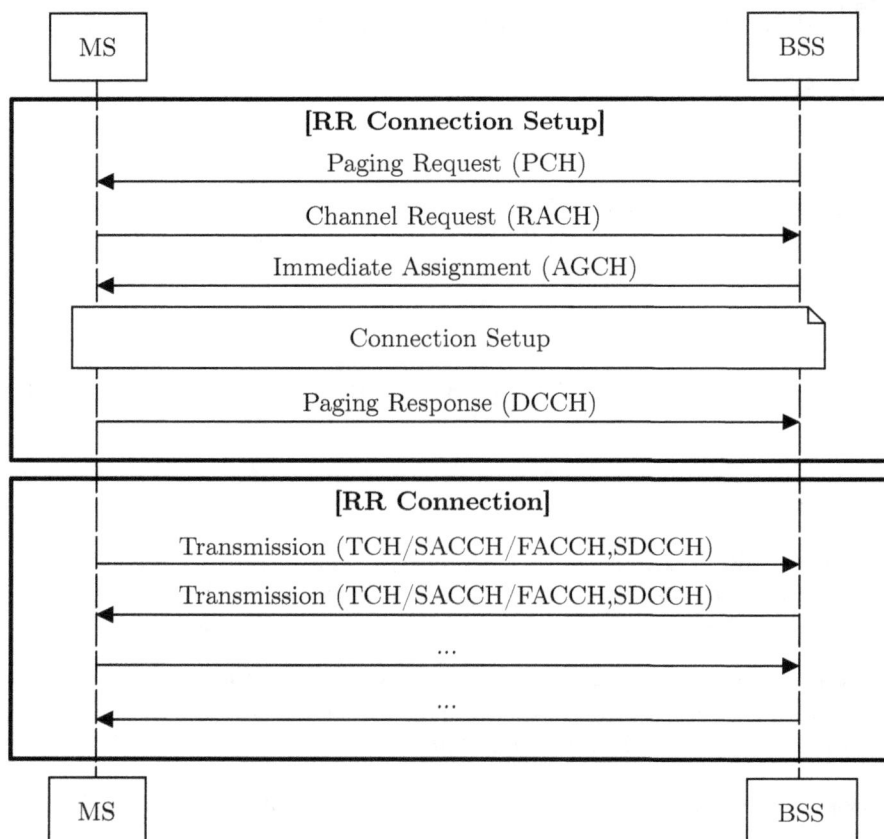

Figure C.1: Sequence diagram for the voice and data connection

In the presented case, the radio connection establishment is initiated by the network (normally on behalf of another subscriber) and terminates on the mobile station. Therefore this kind of connection establishment is denoted as *Mobile Terminated Call (MTC)*.

The paging request on the paging channel informs a specific mobile station about the need to establish a connection and is answered by the mobile station by a channel request message on the random access channel. The base station subsystem then assigns a channel by an immediate assignment message on the access grant channel. After the connection has been set up, the mobile stations answers the paging request by a paging response message on a dedicated control channel.

This procedure establishes a bi-directional dedicated radio resource connection. The following signal transmissions take place on the traffic channel with its slow and fast associated control channels and the slow dedicated control channel.

An arbitrary frequency hopping scheme can be accomplished by assigning a suitable mobile allocation table, which may contain up to 64 frequencies, and setting the hopping sequence number and the mobile allocation index offset to suitable values.

Properties of Transmitted Signals

The transmitted GSM signals are characterized by some important features which are discussed in the following.

Since the signal transmission initiation is based on voice and data transmission, a 26-multiframe is used. Depending on the traffic channel configuration as full-rate or half-rate channel, the 13th and 26th frame have different content.

In case of a full-rate configuration, the 13th frame carries a slow associated control channel and the 26th frame is idle. In case of a half-rate configuration both frames carry a slow associated control channel. In order to prevent empty frames in the wideband signal, a half-rate configuration is therefore preferred.

Furthermore it has to be assured that the *Discontinuous Transmission (DTx)* option is disabled. This method is intended for momentarily powering down the signal transmission when there is no voice input to the mobile station. This approach is very common in GSM systems but can also be disabled by the network. If the option is active, the signal transmission might be halted or interrupted. [79].

The presented signal transmission initiation procedure has the following advantages:

- The signal transmission and therefore the localization is possible during normal mobile phone operation.

- The duration of the wideband signal is not limited.

The drawbacks of the procedure can be summarized as follows:

- The base transceiver station has to support up to 64 frequencies.

- The signaling link is bi-directional, i.e. signal transmission takes place in the uplink and downlink band. Since only a small portion of the downlink band is assigned to the respective GSM service provider or some channels are possibly blocked (e.g. by jamming) there are restrictions on the usable frequencies.

- Several mobile phones may be active at the same time. Therefore the corresponding channel assignments have to be available for separation of the signal sources.

- The cooperation of the user (by answering the call) is necessary for establishing the complete transmission link.

The drawbacks of the voice and data connection pose substantial restrictions for the effective application of the localization system. Therefore, this approach is not preferred for initiating a frequency hopping signal transmission.

C.3 Modified Handover Procedure

The main idea of this approach is the pretense of a non-existing base station and the request to perform a handover to this station. Consequently, the mobile station may transmit a frequency hopping signal.

Transmission Initiation Procedure

For this purpose, the standard handover procedure has to be modified. A sequence diagram for this procedure is depicted in Fig. C.2.

As a first step, the initial *Base Transceiver Station (BTS0)* establishes a traffic channel with the *Mobile Station (MS)*. This channel configuration does not necessarily provide frequency hopping capabilities and can be fully established as described in the previous section (with cooperation of the user) or partially (independent of user activity). A suitable approach for the partial establishment of the channel for measurement purposes is described in [78]. The mobile station is now in dedicated mode.

The mobile station is commanded to perform a non-synchronized handover to another non-existing *Base Transceiver Station (BTS*)* with a supposed traffic channel with frequency hopping. The corresponding hopping parameters and frequencies are communicated to the mobile station in the handover command.

The mobile station suspends the previous channel configuration and starts to send access bursts on the supposed frequency hopping channel. At the same time, the internal timer T3124 is started. The transmission of access bursts is stopped when the mobile station receives a response of the targeted (non-existing) base transceiver station or when the timer T3124 expires.

Since in the 26-multiframe only 24 frames are allocated to a traffic channel and the pre-defined value of the timer is 320 ms, the maximum number of access bursts is

$$N_{max,Handover} = \frac{24}{26} \cdot \frac{320\,\text{ms}}{4.615\,\text{ms}} \approx 64 \qquad \text{(C.1)}$$

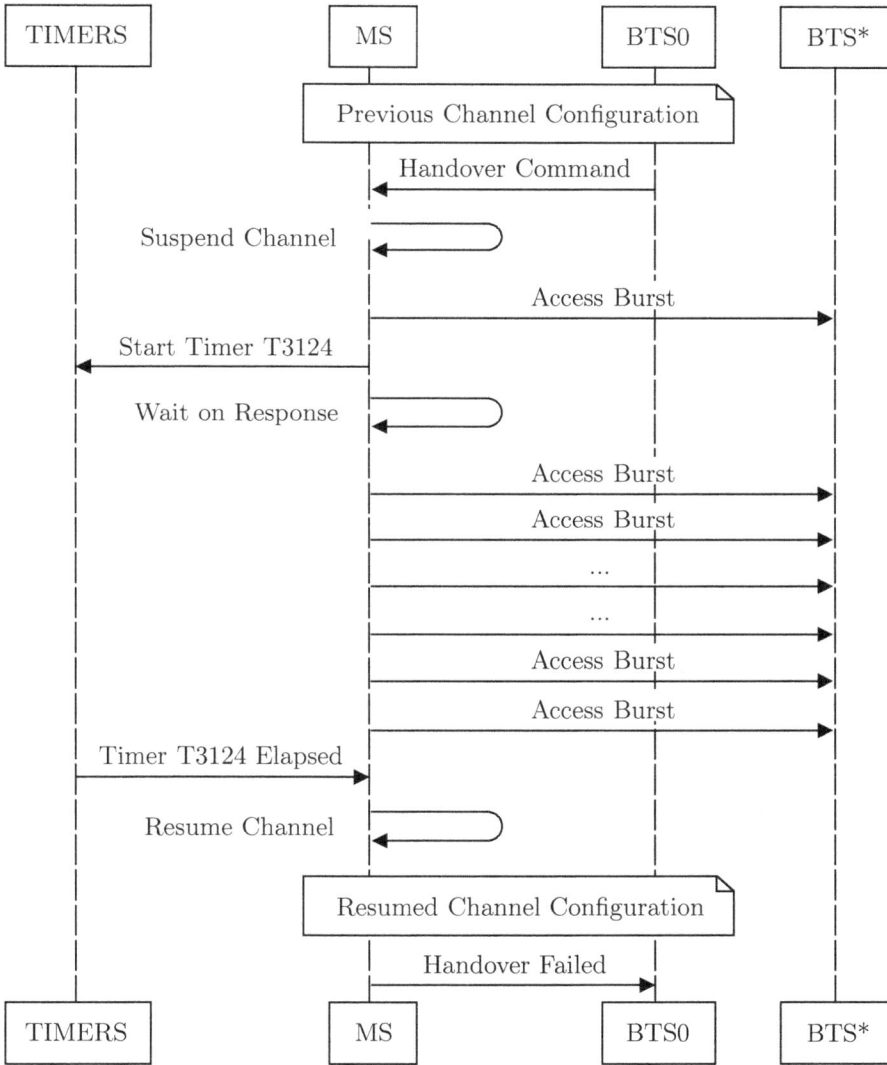

Figure C.2: Sequence diagram for the modified handover procedure

which equals the maximum number of possible frequencies in the mobile allocation table. When the timer T3124 has elapsed, the old channel configuration with the initial base transceiver station is resumed and a handover failed message is reported.

The presented approach is commonly used for localization purposes in the GSM network as described in [4].

Properties of Transmitted Signals

The transmitted GSM signals are characterized by some important features which are discussed in the following.

The described procedure is based on the handover between traffic channels. The signal transmission is therefore basically defined in 24 of the 26 frames of the 26-multiframe. This assumption results in a maximum of 64 transmitted access bursts.

The mobile stations may optionally transmit access bursts on the corresponding slow associated control channels. This possibility has to be taken in account when initiating the signal transmission. [78]

Furthermore, the mobile station is commanded to perform a non-synchronized handover assuming an unknown time offset with regard to the non-existing base transceiver station. Therefore, the access bursts may not be exactly aligned to the original TDMA frame structure.

The presented signal transmission initiation procedure has the following advantages:

- Since no bi-directional signaling link is established, the allocated downlink frequencies have not to be considered. The employed frequencies for signal transmission in the uplink band can be chosen arbitrarily.

- The initial traffic channel can be established without user interaction, i.e. no cooperation of the user is necessary.

The drawbacks of the procedure can be summarized as follows:

- The presented approach exploits the properties of the GSM handover procedure in an originally unintended way. Therefore, modifications of standard procedures are necessary.

- The duration of the wideband signal is limited by the maximum number of transmitted access bursts.

The advantages of the modified handover procedure are essential for the effective application of the localization system and outweigh the drawbacks by far. Therefore, this approach is preferred for initiating a frequency hopping signal transmission.

D Time Difference of Arrival Estimation for Narrowband GSM Signals

In this chapter, time difference of arrival estimation techniques for narrowband GSM signals are presented. These techniques represent the state-of-the-art of time difference of arrival estimation for GSM signals and are the basis of time difference of arrival based GSM mobile phone localization today. The signal acquisition is based on narrowband frontends suited for application in existing GSM networks.

D.1 Narrowband Crosscorrelation Technique

An intuitive approach for estimating the time difference of arrival between two GSM signals is the calculation of the *Crosscorrelation Function (CCF)* between the two burst signals. The crosscorrelation function between two signals $r_{A,ECB}(t)$ and $r_{B,ECB}(t)$ is defined as

$$\text{CCF}(\Delta\tau) = r_{A,ECB}(t) \star r_{B,ECB}(t)\Big|_{\Delta\tau} = \int_{-\infty}^{\infty} r_{A,ECB}^*(t) r_{B,ECB}(t + \Delta\tau)\, \mathrm{d}t \qquad \text{(D.1)}$$

where $\Delta\tau$ denotes the time delay between the two signals and $\star$ represents the crosscorrelation operator.

The time delay that maximizes the absolute value of the crosscorrelation function $\Delta\hat{\tau}_{BA}$ is interpreted as estimation for the actual time difference of arrival, i.e.

$$\Delta\hat{\tau}_{BA} = \text{argmax}_{\Delta\tau}|\text{CCF}(\Delta\tau)| \qquad \text{(D.2)}$$

The shape and the properties of the crosscorrelation function are mainly influenced by the ambiguity function of the transmitted signal. The ambiguity function of narrowband burst signals $s_{Burst,ECB}(t) \star s_{Burst,ECB}(t)$ has been studied e.g. in [1, 80] and is depicted in Fig. D.1.

The maximum of the crosscorrelation function is the optimal estimator for multipath-free scenarios reaching the *Cramér-Rao Lower Bound (CRLB)* i.e. attaining minimum

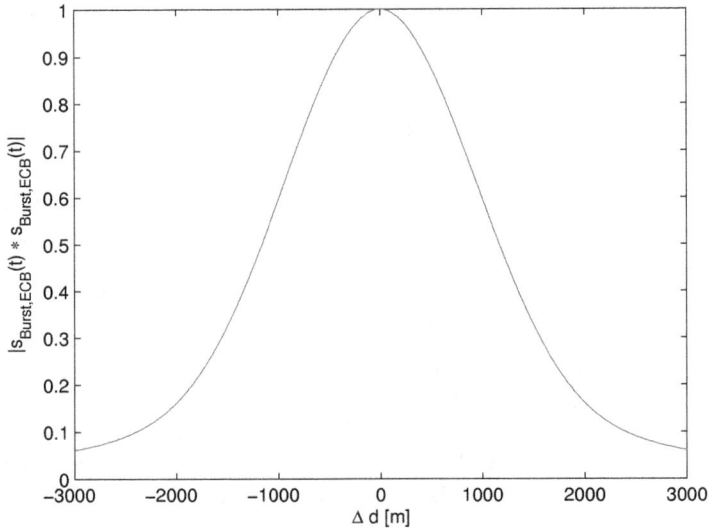

Figure D.1: Ambiguity function of narrowband burst signals

variance of the time difference of arrival estimate. The main drawback of this approach is its performance in multipath scenarios.

The width of the main peak (full width half maximum) is about 2.2 km, i.e. only multipath components exceeding this distance can be separated from the desired line-of-sight component. Any shorter multipath components distort the correlation peak, shift its maximum and therefore lead to severe errors and deviations in the time difference of arrival estimation.

The width of the correlation peak is solely dependent on the signal bandwidth which is a consequence of the Wiener-Khinchine theorem. Since the bandwidth of a single burst signal is constant at about 200 kHz, the width of the main peak is constant as well. The use of longer sequences for crosscorrelation increases the signal energy but not the multipath separability.

A common variant of the presented crosscorrelation between two received signals is the crosscorrelation between the received signal and the corresponding training sequence of the burst in each receiving station separately. The two local delay estimates can then be subtracted in order to yield a resulting delay estimate between the two received signals. The properties of the crosscorrelation function apply accordingly.

In real scenarios, the crosscorrelation of two single burst signals leads to deviations in the delay estimate in the scale of $300 - 400$ m. [81]

D.2 Model-based Estimation Techniques

Within the scientific community considerable efforts have been made in order to improve the accuracy. Therefore, model-based techniques have gained significant interest. The presented approach is based on [82].

Model-based techniques rely on modeling the radio channel impulse response $h(t)$ as a sum of weighted and shifted delta functions according to

$$h(t) = \sum_{i=0}^{I-1} \alpha_i \delta(t - \tau_i) \tag{D.3}$$

with α_i denoting the complex channel coefficients and τ_i the corresponding delays. I is the number of paths.

This model is often transfered to frequency domain with a corresponding model for the frequency response $H(f)$ given as

$$H(f) = \mathcal{F}\{h(t)\} = \sum_{i=0}^{I-1} \alpha_i e^{-j2\pi f \tau_i} \tag{D.4}$$

with $\mathcal{F}\{\cdot\}$ representing the Fourier transform. Consequently, parametric methods for resolving closely spaced frequency components can be employed.

As a first step, the received signals are equalized with respect to the pulse shaping and reception filters. Thus, yielding an estimate for $h(t)$ and $H(f)$, respectively. For this purpose, the training sequence or a local replica of the transmitted signal (which can be obtained by first decoding the received data signal) is used.

Since model-based techniques commonly exhibit poor performance at low signal to noise ratios, a *Singular Value Decomposition (SVD)* is often applied for noise reduction purposes. [82]

Finally, the unknown parameters are estimated using model-based techniques such as

- *Modified Least Squares Prony (MLS-Prony)* [82]
- *Root Multiple Signal Classification (Root-MUSIC)* [82]
- *Maximum Likelihood (ML) Estimation* [83]
- *Signal Eigen Vector Decomposition* [84]

The smallest value of the path delay estimates $\hat{\tau}_i$ is then interpreted as actual time delay estimate of the respective signal, i.e.

$$\hat{\tau} = \min_i \hat{\tau}_i \tag{D.5}$$

This procedure is performed in all receiving stations separately. The time delay estimates $\hat{\tau}_A$ and $\hat{\tau}_B$ for receiving station A and B can then finally be subtracted yielding an estimate for the time difference of arrival between the received signals, i.e.

$$\Delta\hat{\tau}_{BA} = \hat{\tau}_B - \hat{\tau}_A \tag{D.6}$$

The drawbacks of these techniques are their need for high signal to noise ratios and the requirement of a-priori knowledge of the model order I. Therefore, the model order has to be estimated by suitable techniques or heuristic approaches beforehand, which often leads to unreliable results.

However, the performance of model-based techniques is substantially improved compared to the narrowband crosscorrelation technique. The accuracy for delay estimation is shown to be in the scale of $50 - 200\,\text{m}$. [81]

D.3 Incoherent Integration with Multipath Rejection Technique

The *Incoherent Integration (ICI)* technique is based on multiple evaluations of narrowband crosscorrelation functions and incoherent superposition of the correlation results. The presented procedures are described in more detail in [59, 60].

As a first step, an estimate of the radio channel impulse response $h_n(t)$ is obtained by calculating the narrowband crosscorrelation function between the received burst signal n and the corresponding training sequence. For each estimate, the corresponding *Channel Power Profile (CPP)* is calculated as

$$\text{CPP}_n(t) = |h_n(t)|^2 \tag{D.7}$$

The main idea of the incoherent integration approach is the incoherent summation over multiple channel power profiles given as

$$\text{ICI}(t) = \sum_{n=0}^{N-1} \gamma_n |h_n(t)|^2 \tag{D.8}$$

with γ_n denoting weights based on the estimated quality of each $h_n(t)$.

The time delay that maximizes the incoherent integration function $\text{ICI}(t)$ is then interpreted as estimation for the actual time delay, i.e.

$$\hat{\tau} = \text{argmax}_t\ \text{ICI}(t) \tag{D.9}$$

Since the technique exploits different kinds of diversity such as frequency diversity (in form of optional frequency hopping), time diversity (in form of a possibly moving mobile phone) and space diversity (using possibly multiple antennas), a substantial improvement in estimation accuracy is expected.

The *Incoherent Integration with Multipath Rejection (ICI-MPR)* technique is characterized by an additional multipath rejection procedure.

For this purpose, subsets of randomly chosen channel power profiles are formed and the incoherent integration is performed within each subset. Consequently, partial sums for each subset may be obtained as

$$\text{ICI}_m(t) = \sum_{i \in \Omega_m} \gamma_i |h_i(t)|^2 \tag{D.10}$$

with m denoting the subset index and Ω_m the subset containing the respective burst indices.

A time delay estimate for each subset can then be calculated as

$$\hat{\tau}_m = \text{argmax}_t \ \text{ICI}_m(t) \tag{D.11}$$

For assumed *Line-of-Sight (LOS)* scenarios, the resulting time delay estimate is finally obtained as mean value over the subset estimates, i.e.

$$\hat{\tau}_{LOS} = \text{mean}(\hat{\tau}_m) \tag{D.12}$$

In case of assumed *Non Line-of-Sight (NLOS)* scenarios, the standard deviation of the subset estimates is additionally subtracted, i.e.

$$\hat{\tau}_{NLOS} = \text{mean}(\hat{\tau}_m) - \text{std}(\hat{\tau}_m) \tag{D.13}$$

The decision, whether a strong line-of-sight component is present or not, is based on the standard deviation of all subset estimates compared to the standard deviation of a noise scenario.

The main idea of the multipath rejection approach is the identification of LOS and NLOS scenarios based on the assumption that some subsets mostly comprise burst signals with no or minor multipath distortions.

The incoherent integration with multipath rejection procedure is performed in all receiving stations separately. The time delay estimates $\hat{\tau}_A$ and $\hat{\tau}_B$ for receiving station A and B can then finally be subtracted yielding an estimate for the time difference of arrival between the received signals, i.e.

$$\Delta\hat{\tau}_{BA} = \hat{\tau}_B - \hat{\tau}_A \tag{D.14}$$

In summary, the presented techniques provide a substantially improved accuracy compared to the narrowband crosscorrelation and model-based estimation techniques. The accuracy is shown to be in the scale of $50 - 100\,\text{m}$. [59]

D.4 Successive Cancellation Technique

The successive cancellation technique is an iterative approach which interprets estimation results from a previous cycle as interference and removes the respective components of the signal. The presented procedures are described in [85, 86].

The N received burst signals are modeled as

$$r^{(n)}(t) = \sum_{i=0}^{I-1} \alpha_i^{(n)} s(t - \tau_i) + n^{(n)}(t) \tag{D.15}$$

with $\alpha_i^{(n)}$ denoting the complex channel attenuations and τ_i the corresponding time delays of multipath components. $n^{(n)}(t)$ represents a Gaussian noise process with known autocorrelation function.

In the sampled domain, this model can be expressed as

$$r^{(n)}[k] = \sum_{i=0}^{I-1} \alpha_i^{(n)} s(kT_s - \tau_i) + n^{(n)}(kT_s) \tag{D.16}$$

with T_s denoting the sampling period and $0 \leq k \leq K - 1$.

The fundamental estimation techniques are the narrowband crosscorrelation technique or the model-based MUSIC approach. In addition, an incoherent integration over multiple measurements is integrated in the estimation procedure. In the following, the correlation based approach is described.

The conventional crosscorrelation function with incoherent integration in vector notation can be expressed as

$$c(\tau) = \frac{1}{N} \sum_{n=0}^{N-1} |\mathbf{s}(\tau)^{\mathbf{H}} \cdot \mathbf{r}^{(n)}|^2 \tag{D.17}$$

with

$$\mathbf{s}(\tau) = [s(0 - \tau), s(T_s - \tau), s(2T_s - \tau), \ldots, s((K-1)T_s - \tau)] \tag{D.18}$$

and

$$\mathbf{r}^{(n)} = [r^{(n)}[0], r^{(n)}[1], r^{(n)}[2], \ldots, r^{(n)}[K-1]] \tag{D.19}$$

This expression is modified using the correlation matrix $\mathbf{C_K}$ of the impairment process $n(kT_s)$ which accounts for correlations within the process. The modified crosscorrelation function with incoherent integration can be expressed as

$$c(\tau) = \frac{1}{N} \sum_{n=0}^{N-1} |(\mathbf{C_K^{-1}} \cdot \mathbf{s}(\tau))^{\mathbf{H}} \cdot \mathbf{r}^{(n)}|^2 \tag{D.20}$$

The successive cancellation approach is based on recomputing the correlation matrix $\mathbf{C_K}^{(m+1)}$ in iteration cycle $m+1$ according to

$$\mathbf{C_K}^{(m+1)} = \mathbf{C_K}^{(m)} + \mathbf{S}^{(m)} \cdot \mathbf{P} \cdot \mathbf{S}^{(m)\mathbf{H}} \tag{D.21}$$

with $\mathbf{P}$ representing the fixed correlation matrix of the radio channel and $\mathbf{S}^{(m)}$ containing the signal components at the estimated time delays from the previous iterations, i.e.

$$\mathbf{S}^{(m)} = \begin{bmatrix} s(0 - \hat{\tau}_0^{(m)}) & \cdots & s(0 - \hat{\tau}_{I-1}^{(m)}) \\ s(T_s - \hat{\tau}_0^{(m)}) & \cdots & s(T_s - \hat{\tau}_{I-1}^{(m)}) \\ \vdots & & \vdots \\ s((K-1)T_s - \hat{\tau}_0^{(m)}) & \cdots & s((K-1)T_s - \hat{\tau}_{I-1}^{(m)}) \end{bmatrix} \tag{D.22}$$

In the first cycle, $\mathbf{C_K}^{(0)}$ is supposed to be an identity matrix. The total number of iterations M may either be a fixed number or may be adapted based on the maximum correlation amplitude after each iteration.

This procedure effectively treats the detected signal vectors up to this iteration as noise and suppresses the signal components at these delays in the correlation output. Consequently, other signal components may be detected. After the last iteration, all estimated time delays are tabulated and the earliest delay is chosen as actual time delay estimate, i.e.

$$\hat{\tau} = \min_i \hat{\tau}_i \tag{D.23}$$

The successive cancellation procedure is performed in all receiving stations separately. The time delay estimates $\hat{\tau}_A$ and $\hat{\tau}_B$ for receiving station A and B can then finally be subtracted yielding an estimate for the time difference of arrival between the received signals, i.e.

$$\Delta\hat{\tau}_{BA} = \hat{\tau}_B - \hat{\tau}_A \tag{D.24}$$

In summary, the presented technique provides a substantially improved accuracy compared to the narrowband crosscorrelation and model-based estimation techniques. The accuracy is shown to be in the scale of $50 - 100\,\text{m}$. [86]

D.5 Further Applicable Techniques

Time difference of arrival estimation techniques have been studied within the scientific community extensively for a variety of applications. Some of these techniques might also be applicable for GSM signals and are discussed in the following.

Generalized Crosscorrelation Techniques

The *Generalized Crosscorrelation (GCC)* technique is based on filtering the received signals prior to the crosscorrelation operation. The filter characteristics may be adapted to meet specific criteria such as noise performance. [87]

The generalized crosscorrelation technique is a generalization of the narrowband crosscorrelation technique and does not address the insufficient multipath separability of narrowband signals.

Therefore, only minor improvements compared to the narrowband crosscorrelation technique can be expected.

Generalized Model-based Techniques

Model-based techniques have been adapted for different time difference of arrival estimation purposes. In addition to the presented model-based techniques, a multitude of variations exists. As an example, a variation of the MUSIC approach for time difference of arrival estimation is described in [88].

Furthermore, alternative techniques for the joint estimation of delay, azimuth and elevation, which normally are intended for use in sensor arrays, could be adapted. Among others, the ESPRIT [89, 90], Unitary-ESPRIT [91], JADE [92] and SAGE [93] algorithms may be of interest.

All these techniques share the same advantages and disadvantages as the presented model-based techniques for GSM signals. Therefore, only minor improvements with respect to time difference of arrival estimation can be expected.

Spectral Estimation Techniques for Transfer Function Analysis

As described in the section on model-based estimation techniques for GSM signals, the estimation of time delay may be interpreted as spectral estimation of the radio channel transfer function $H(f)$. Therefore, established methods for spectral analysis may be applied.

Common techniques for spectral estimation comprise e.g. the Fourier-based periodogram and its variants such as the Bartlett and Welch procedure [94]. Furthermore, spectral estimation according to Capon's minimum variance method can be applied for time difference of arrival estimation [95].

The multipath resolvability of the Fourier-based techniques is limited by the bandwidth of the narrowband signals. Therefore, their performance is expected to be comparable to the narrowband crosscorrelation technique. Capon's minimum variance method has shown to be comparable to the successive cancellation technique [85, 86].

Channel Estimation Techniques for Communication Systems

In communication systems, channel estimation is a common prerequisite for the equalization of received signals. The equalization compensates for the *Inter Symbol Interference (ISI)* which is introduced by multipath propagation. [37]

The channel estimates intended for equalization may also be applied for time difference of arrival estimation. Therefore, these techniques might also be of interest.

Channel estimation techniques may be categorized into blind techniques which are based on the application of reference data and non-blind techniques without the application of reference data. As an example, the narrowband crosscorrelation approach with training sequence can be interpreted as a non-blind technique.

For GSM systems, different algorithms have been studied e.g. in [96], [97] and [98]. The performance of the techniques is commonly expressed in terms of *Bit Error Rate (BER)* curves for information transmission. Therefore, the time difference of arrival estimation performance is hardly measurable.

Since the techniques show comparable results as the narrowband crosscorrelation approach with training sequence, no substantial improvements with respect to the multipath separability are expected.

Synchronization and Timing Recovery Techniques for Communication Systems

In communication systems, the carrier synchronization and symbol timing recovery are essential for the demodulation of the communication signals. The carrier phase synchronization and symbol timing may also be applied for time difference of arrival estimation.

The applicability of either approach depends on the corresponding symbol propagation length $c_0 T_{Symbol}$ and the wavelength c_0/f_c due to the inherent ambiguities of successive symbols and carrier wave periods, respectively.

In GSM systems, only the symbol timing recovery with a symbol length of approximately 1.1 km is of interest. Different algorithms for timing recovery in GSM systems have been studied e.g. in [99], [100], [101] and [102].

The performance of these techniques is commonly investigated for AWGN scenarios only. Since the performance in multipath scenarios is decisive for the overall estimation accuracy, no reliable performance evaluation based on the mentioned publications is possible. However, conducted computer simulations indicate that the multipath performance is comparable to the performance of the narrowband crosscorrelation technique.

D.6 Comparative Summary

In the following, the properties of the presented techniques for time difference of arrival estimation for narrowband GSM signals are summarized. The advantages and disadvantages are briefly indicated in Tab. D.1.

Estimation Technique	Resolvability	Reliability	Accuracy
Narrowband Crosscorrelation	−	+	$300 - 400\,\mathrm{m}$
Model-based Techniques	+	−	$50 - 200\,\mathrm{m}$
Incoherent Integration	−	+	$50 - 100\,\mathrm{m}$
Successive Cancellation	−	0	$50 - 100\,\mathrm{m}$

Table D.1: Summary of estimation techniques for narrowband GSM signals

In the table, the resolvability refers to the ability of the techniques to identify individual multipath components. The reliability indicates whether a-priori assumptions influence the estimation results. Furthermore, the robustness of the results is characterized by this measure. The accuracy denotes the deviation of the estimates relative to the desired time difference of arrival value.

The further applicable techniques can be interpreted as generalizations or deviations of the previously presented techniques. Moreover, a different perspective on time difference of arrival estimation from the point of communication systems is provided. The properties of these techniques can be assumed to be similar to the previously presented techniques.

All techniques for time difference of arrival estimation of narrowband GSM signals are essentially based on the processing of single burst signals with an approximate bandwidth of $200\,\mathrm{kHz}$ for each burst. Some techniques rely on the advantageous combination of estimation results in order to improve the estimation performance. However, there are fundamental limits on the attainable accuracy and resolution as described in [62, 63].

E Localization System Architecture for Search and Rescue Scenarios

The presented techniques for time difference of arrival estimation provide a substantially improved performance compared to state-of-the-art techniques. Therefore, novel applications with more challenging requirements regarding localization accuracy and resolution may be realized.

In the following, the application of the system in search and rescue scenarios is focused and a preferred system architecture is described. The overall structure of the proposed localization system is depicted in Fig. E.1.

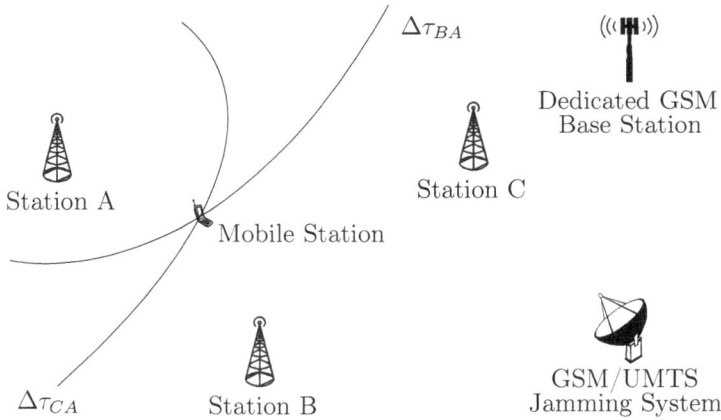

Figure E.1: Structure of the proposed localization system

The mobile phone, or mobile station, is supposed to be carried by the person that is be located. The mobile phone may be logged on to an existing GSM base station of a network provider or may have lost its connection depending on the scale of the natural disaster.

In the following, it is assumed that the mobile phone is either in an idle state or is scanning for existing base stations. The case of an active voice and data connection may be advantageous but requires the knowledge of the allocated channel resources of the mobile phone which is to be localized.

The receiving stations A, B and C operate passively and determine the *Time Differences of Arrival (TDOAs)* of the GSM signal between corresponding pairs of receiving stations. The time difference of arrival of the signal between receiving station A and B is denoted $TDOA_{BA}$. The time difference of arrival of the signal between receiving station A and C is denoted $TDOA_{CA}$. The location of the mobile station can then be determined using multilateration techniques.

For a two-dimensional localization, at least three receiving stations are necessary. In a three-dimensional localization scenario, at least four receiving stations are required. Furthermore it is essential that the geometric setup, i.e. the locations of the receiving stations is chosen advantageously in terms of a low *Geometric Dilution of Precision (GDOP)* value.

For the determination of the time difference of arrival values, the receiving stations have to be synchronized. The synchronization signals may be derived from a *Global Navigation Satellite System (GNSS)*, a wired synchronization network or a wireless synchronization network. The synchronization signals may then serve as reference e.g. for *Phase Locked Loops (PLLs)*.

For the application of the multilateration technique, the location of the receiving stations has to be known. If the receiving stations are equipped with GNSS receivers, the location of the receiving stations can be determined in form of global coordinates. An alternative approach is the application of a local positioning system establishing a local coordinate system. Otherwise the locations of the receiving stations have to be determined manually by the rescue personnel.

The dedicated GSM base station serves the purpose of controlling the logged in mobile phones. In particular, the mobile phones have to be separated and be initiated to start a signal transmission. The base station may be designed to operate in the E-GSM 900 frequency band which provides a bandwidth of 35 MHz for localization and a higher penetration of covering debris material due to the lower carrier frequencies compared to DCS 1800 or PCS 1900. However, an operation in the DCS 1800 or PCS 1900 band may provide higher bandwidths for localization of 75 MHz or 60 MHz, respectively.

The GSM and UMTS radio frequency jamming system is essential if the mobile phone is connected to an existing base station of a network provider. Since the mobile phone has to be controlled by the dedicated GSM base station, all connections to existing GSM and UMTS networks have to be disconnected. The application of a UMTS jamming system enables even the localization of UMTS mobile phones, since GSM is inherently supported by UMTS mobile phones and used as fall back solution if no UMTS network is available.

The control of the components and calculation of the time difference of arrival values is performed on a central computing device, such as a personal computer or server. The localization results may then be projected on a map or combined with further localization results obtained e.g. by search dogs, acoustic localization or search probes.

References

[1] C. Drane, M. Macnaughtan, and C. Scott, "Positioning GSM Telephones," *IEEE Communications Magazine*, vol. 36, no. 4, pp. 46–54, 1998.

[2] "Commission Recommendation on the Processing of Caller Location Information in Electronic Communication Networks for the Purpose of Location-Enhanced Emergency Call Services," Official Journal of the European Union, July 25, 2003, Document Number C(2003) 2657.

[3] "Framework for Next Generation 911 Deployment, Notice of Inquiry," Federal Communications Commission, December 21, 2010, PS Docket No. 10-255.

[4] *3GPP TS 03.71 V8.9.0*, 3rd Generation Partnership Project Std.

[5] M. Tayal, "Location Services in the GSM and UMTS Networks," in *Proceedings IEEE International Conference on Personal Wireless Communications*, 2005, pp. 373–378.

[6] D.-B. Lin and R.-T. Juang, "Mobile Location Estimation Based on Differences of Signal Attenuations for GSM Systems," *IEEE Transactions on Vehicular Technology*, vol. 54, no. 4, pp. 1447–1454, 2005.

[7] M. A. Spirito, S. Poykko, and O. Knuuttila, "Experimental Performance of Methods to Estimate the Location of Legacy Handsets in GSM," in *Proceedings IEEE Vehicular Technology Conference*, vol. 4, 2001, pp. 2716–2720.

[8] *3GPP TR 45.811 V6.0.0*, 3rd Generation Partnership Project Std.

[9] M. A. Spirito and A. G. Mattioli, "On the Hyperbolic Positioning of GSM Mobile Stations," in *Proceedings URSI International Symposium on Signals, Systems and Electronics*, 1998, pp. 173–177.

[10] M. A. Spirito, "Further Results on GSM Mobile Station Location," *IET Electronics Letters*, vol. 35, no. 11, pp. 867–869, 1999.

[11] V. Ruutu, M. Alanen, G. Gunnarsson, T. Rantalainen, and V.-M. Teittinen, "Mobile Phone Location in Dedicated and Idle Modes," in *Proceedings IEEE International Symposium on Personal, Indoor and Mobile Radio Communications*, vol. 1, 1998, pp. 456–460.

[12] S. Anderson, B. Hagerman, H. Dam, U. Forssen, J. Karlsson, F. Kronestedt, S. Mazur, and K. J. Molnar, "Adaptive Antennas for GSM and TDMA Systems," *IEEE Personal Communications*, vol. 6, no. 3, pp. 74–86, 1999.

[13] M. C. Wells, "Increasing the Capacity of GSM Cellular Radio using Adaptive Antennas," *IEE Proceedings Communications*, vol. 143, no. 5, pp. 304–310, 1996.

[14] S. Sakagami, S. Aoyama, K. Kuboi, S. Shirota, and A. Akeyama, "Vehicle Position Estimates by Multibeam Antennas in Multipath Environments," *IEEE Transactions on Vehicular Technology*, vol. 41, no. 1, pp. 63–68, 1992.

[15] F. Cesbron and R. Arnott, "Locating GSM Mobiles using Antenna Array," *IET Electronics Letters*, vol. 34, no. 16, pp. 1539–1540, 1998.

[16] N. Deligiannis and S. Louvros, "Hybrid TOA–AOA Location Positioning Techniques in GSM Networks," *Springer Wireless Personal Communications*, vol. 54, pp. 321–348, 2010.

[17] A. L. Kabir, R. Saha, M. A. Khan, and M. M. Sohul, "Locating Mobile Station Using Joint TOA/AOA," in *Proceedings International Conference on Ubiquitous Information Technologies & Applications*, 2009, pp. 1–6.

[18] B. Denby, Y. Oussar, I. Ahriz, and G. Dreyfus, "High-Performance Indoor Localization with Full-Band GSM Fingerprints," in *Proceedings IEEE International Conference on Communications Workshops*, 2009, pp. 1–5.

[19] M. Ibrahim and M. Youssef, "CellSense: An Accurate Energy-Efficient GSM Positioning System," *IEEE Transactions on Vehicular Technology*, vol. 61, no. 1, pp. 286–296, 2012.

[20] Z. Salcic and E. Chan, "Mobile Station Positioning Using GSM Cellular Phone and Artificial Neural Networks," *Springer Wireless Personal Communications*, vol. 14, pp. 235–254, 2000.

[21] J. F. Bull, "Wireless Geolocation," *IEEE Vehicular Technology Magazine*, vol. 4, no. 4, pp. 45–53, 2009.

[22] M. I. Silventoinen and T. Rantalainen, "Mobile Station Emergency Locating in GSM," in *Proceedings IEEE International Conference on Personal Wireless Communications*, 1996, pp. 232–238.

[23] T. Wigren, M. Anderson, and A. Kangas, "Emergency Call Delivery Standards Impair Cellular Positioning Accuracy," in *Proceedings IEEE International Conference on Communications*, 2010, pp. 1–6.

[24] T. Wigren, "A Polygon to Ellipse Transformation Enabling Fingerprinting and Emergency Localization in GSM," *IEEE Transactions on Vehicular Technology*, vol. 60, no. 4, pp. 1971–1976, 2011.

[25] D. Tassetto, E. H. Fazli, and M. Werner, "A Novel Hybrid Algorithm for Passive Localization of Victims in Emergency Situations," in *Proceedings Advanced Satellite Mobile Systems Conference*, 2008, pp. 320–327.

[26] P. Morgand, A. Ferreol, R. Sarkis, C. Craeye, and C. Oestges, "Detection and Location of People in Emergency Situations through their PMR or GSM/UMTS Phones," in *Proceedings European Wireless Technology Conference*, 2010, pp. 185–188.

[27] HEPKIE System. ResQU AB, Sweden. [Online]. Available: http://hepkie.com

[28] J. G. Proakis and D. G. Manolakis, *Digital Signal Processing*, 3rd ed.   Prentice Hall, 1996.

[29] C. Mensing and S. Plass, "Positioning Algorithms for Cellular Networks Using TDOA," in *Proceedings IEEE International Conference on Acoustics, Speech and Signal Processing*, vol. 4, 2006.

[30] G. Mao and B. Fidan, *Localization Algorithms and Strategies for Wireless Sensor Networks*.   Idea Group Reference, 2009.

[31] K. J. Krizman, T. E. Biedka, and T. S. Rappaport, "Wireless Position Location: Fundamentals, Implementation Strategies, and Sources of Error," in *Proceedings IEEE Vehicular Technology Conference*, vol. 2, 1997, pp. 919–923.

[32] N. Levanon, "Lowest GDOP in 2-D scenarios," *IEE Proceedings Radar, Sonar and Navigation*, vol. 147, no. 3, pp. 149–155, 2000.

[33] D.-H. Shin and T.-K. Sung, "Analysis of Positioning Errors in Radionavigation Systems," in *Proceedings IEEE Intelligent Transportation Systems Conference*, 2001, pp. 156–159.

[34] J. G. Proakis, *Digital Communications*, 4th ed.   McGraw-Hill, 2001.

[35] J. Huber, "Nachrichtenübertragung," Lecture Notes, Lehrstuhl für Informationsübertragung, Universität Erlangen-Nürnberg, Germany, October 2006, in German Language.

[36] J. Eberspächer, H.-J. Vögel, C. Bettstetter, and C. Hartmann, *GSM – Architecture, Protocols and Services*, 3rd ed.   John Wiley & Sons, 2009.

[37] T. S. Rappaport, *Wireless Communications – Principles and Practice*, 2nd ed.   Prentice Hall, 2002.

[38] P. Laurent, "Exact and Approximate Construction of Digital Phase Modulations by Superposition of Amplitude Modulated Pulses (AMP)," *IEEE Transactions on Communications*, vol. 34, no. 2, pp. 150–160, 1986.

[39] *3GPP TS 45.004 V9.1.0*, 3rd Generation Partnership Project Std.

[40] *3GPP TS 45.002 V9.3.0*, 3rd Generation Partnership Project Std.

[41] *3GPP TS 45.001 V9.2.0*, 3rd Generation Partnership Project Std.

[42] *3GPP TS 45.005 V9.3.0*, 3rd Generation Partnership Project Std.

[43] M. Hook, C. Johansson, and H. Olofsson, "Frequency Diversity Gain in Indoor GSM Systems," in *Proceedings IEEE Vehicular Technology Conference*, vol. 1, 1996, pp. 316–320.

[44] T. T. Nielsen and J. Wigard, *Performance Enhancements in a Frequency Hopping GSM Network*. Kluwer Academic Publishers, 2000.

[45] S. M. Kay, *Fundamentals of Statistical Signal Processing - Estimation Theory*. Prentice Hall, 1993, vol. I.

[46] M. Pichler, S. Schwarzer, A. Stelzer, and M. Vossiek, "Multi-Channel Distance Measurement With IEEE 802.15.4 (ZigBee) Devices," *IEEE Journal of Selected Topics in Signal Processing*, vol. 3, no. 5, pp. 845–859, 2009.

[47] S. Schwarzer, M. Vossiek, M. Pichler, and A. Stelzer, "Precise Distance Measurement with IEEE 802.15.4 (ZigBee) Devices," in *Proceedings IEEE Radio and Wireless Symposium*, 2008, pp. 779–782.

[48] M. Pichler, S. Schwarzer, A. Stelzer, and M. Vossiek, "Positioning with Moving IEEE 802.15.4 (ZigBee) Transponders," in *Proceedings IEEE International Microwave Workshop on Wireless Sensing, Local Positioning, and RFID*, 2009, pp. 1–4.

[49] A. J. Wilkinson, R. T. Lord, and M. R. Inggs, "Stepped-Frequency Processing by Reconstruction of Target Reflectivity Spectrum," in *Proceedings South African Symposium on Communications and Signal Processing*, 1998, pp. 101–104.

[50] D. E. Maron, "Frequency-Jumped Burst Waveforms with Stretch Processing," in *Proceedings IEEE International Radar Conference*, 1990, pp. 274–279.

[51] R. T. Lord and M. R. Inggs, "High Resolution SAR Processing Using Stepped-Frequencies," in *Proceedings IEEE International Symposium on Geoscience and Remote Sensing*, vol. 1, 1997, pp. 490–492.

[52] ——, "High Range Resolution Radar Using Narrowband Linear Chirps Offset in Frequency," in *Proceedings South African Symposium on Communications and Signal Processing*, 1997, pp. 9–12.

[53] W. Nel, J. Tait, R. Lord, and A. Wilkinson, "The Use of a Frequency Domain Stepped Frequency Technique to Obtain High Range Resolution on the CSIR X-Band SAR System," in *Proceedings IEEE Africon Conference*, vol. 1, 2002, pp. 327–332.

[54] M. A. Richards, *Fundamentals of Radar Signal Processing*. McGraw-Hill, 2005.

[55] M. C. Walden and R. D. Pollard, "On the Processing Gain and Pulse Compression Ratio of Frequency Hopping Spread Spectrum Waveforms," in *Proceedings National Telesystems Conference on Commercial Applications and Dual-Use Technology*, 1993, pp. 215–219.

[56] J.-R. Ohm and H. Lüke, *Signalübertragung*, 9th ed. Springer Verlag, 2005, in German Language.

[57] R. J. Freund, W. J. Wilson, and P. Sa, *Regression Analysis: Statistical Modeling of a Response Variable*, 2nd ed. Academic Press, 2006.

[58] A. Ekstrøm and J. Mikkelsen, *GSMsim*. Aalborg Universitetsforlag, 1997.

[59] S. Fischer, H. Grubeck, A. Kangas, H. Koorapaty, E. Larsson, and P. Lundqvist, "Time of Arrival Estimation of Narrowband TDMA Signals for Mobile Positioning," in *Proceedings IEEE International Symposium on Personal, Indoor and Mobile Radio Communications*, vol. 1, 1998, pp. 451–455.

[60] S. Fischer and A. Kangas, "Time-of-Arrival Estimation for E-OTD Location in GERAN," in *Proceedings IEEE International Symposium on Personal, Indoor and Mobile Radio Communications*, vol. 2, 2001.

[61] C. E. Cook and M. Bernfeld, *Radar Signals – An Introduction to Theory and Application*. Academic Press, 1967.

[62] A. Weiss and E. Weinstein, "Fundamental Limitations in Passive Time Delay Estimation – Part I: Narrow-Band Systems," *IEEE Transactions on Acoustics, Speech, and Signal Processing*, vol. 31, no. 2, pp. 472–486, 1983.

[63] E. Weinstein and A. Weiss, "Fundamental Limitations in Passive Time Delay Estimation – Part II: Wide-Band Systems," *IEEE Transactions on Acoustics, Speech, and Signal Processing*, vol. 32, no. 5, pp. 1064–1078, 1984.

[64] A. Loke and F. Ali, "Direct Conversion Radio for Digital Mobile Phones – Design Issues, Status, and Trends," *IEEE Transactions on Microwave Theory and Techniques*, vol. 50, no. 11, pp. 2422–2435, 2002.

[65] H. Saarnisaari, "Some Design Aspects of Mobile Local Positioning Systems," in *Proceedings IEEE/ION Position Location and Navigation Symposium*, 2004, pp. 300–309.

[66] Local Positioning Radar. Symeo GmbH, Germany. [Online]. Available: http://www.symeo.com

[67] S. Roehr, M. Vossiek, and P. Gulden, "Method for High Precision Radar Distance Measurement and Synchronization of Wireless Units," in *Proceedings IEEE International Microwave Symposium*, 2007, pp. 1315–1318.

[68] S. Roehr, P. Gulden, and M. Vossiek, "Precise Distance and Velocity Measurement for Real Time Locating in Multipath Environments Using a Frequency-Modulated Continuous-Wave Secondary Radar Approach," *IEEE Transactions on Microwave Theory and Techniques*, vol. 56, no. 10, pp. 2329–2339, 2008.

[69] J. Brendel, "Simulation und Entwurf eines Synchronisationsmoduls für ein laufzeitbasiertes Ortungssystem," Intermediate Thesis, Lehrstuhl für Technische Elektronik, Universität Erlangen-Nürnberg, Germany, 2010, in German Language.

[70] ComBlock Modules. Mobile Satellite Services Inc., USA. [Online]. Available: http://www.comblock.com

[71] Pico-ITX Single Board Computer. Kontron AG, Germany. [Online]. Available: www.kontron.com

[72] MATLAB Software Package. MathWorks Inc., USA. [Online]. Available: http://www.mathworks.com

[73] S. Erhardt, "Implementierung und Evaluierung eines drahtlosen TDOA-Ortungssystems," Bachelor Thesis, Lehrstuhl für Technische Elektronik, Universität Erlangen-Nürnberg, Germany, 2011, in German Language.

[74] N.-S. Seong and S.-O. Park, "Clock Offsets in TDOA Localization," *Springer Ubiquitous Computing Systems Lecture Notes in Computer Science*, vol. 4239/2006, pp. 111–118, 2006.

[75] J. Falk, P. Handel, and M. Jansson, "Multisource Time Delay Estimation Subject to Receiver Frequency Errors," in *Proceedings IEEE Asilomar Conference on Signals, Systems and Computers*, vol. 1, 2003, pp. 1156–1160.

[76] N. Shah, T. Kamakaris, U. Tureli, and M. Buddhikot, "Wideband Spectrum Probe for Distributed Measurements in Cellular Band," in *Proceedings ACM International Workshop on Technology and Policy for Accessing Spectrum*, 2006.

[77] S. Kasera, N. Narang, and A. P. Priyanka, *2.5G Mobile Networks: GPRS and EDGE*. Tata McGraw-Hill, 2008.

[78] C. Meier, "Aufbau einer GSM-Zelle zur Emulation eines GSM-Netzwerkes," Diploma Thesis, Lehrstuhl für Technische Elektronik, Universität Erlangen-Nürnberg, Germany, 2009, in German Language.

[79] *3GPP TS 46.031 V9.0.0*, 3rd Generation Partnership Project Std.

[80] Z. Xinghao, T. Ran, and W. Yue, "Analytical Expression of GSM Signal Ambiguity Function," in *Proceedings IEEE International Conference on Signal Processing*, 2008, pp. 2279–2283.

[81] C.-R. Comsa, J. Luo, A. Haimovich, and S. Schwartz, "Wireless Localization using Time Difference of Arrival in Narrow-Band Multipath Systems," in *Proceedings International Symposium on Signals, Circuits and Systems*, vol. 2, 2007, pp. 1–4.

[82] J. Winter and C. Wengerter, "High Resolution Estimation of the Time of Arrival for GSM Location," in *Proceedings IEEE Vehicular Technology Conference*, vol. 2, 2000, pp. 1343–1347.

[83] L. Krasny and H. Koorapaty, "Performance of ML Estimators for Time of Arrival Estimation," in *Proceedings IEEE Vehicular Technology Conference*, vol. 4, 1999, pp. 1972–1976.

[84] ——, "Performance of Time of Arrival Estimation Based on Signal Eigen Vectors," in *Proceedings IEEE International Symposium on Personal, Indoor and Mobile Radio Communications*, vol. 2, 2000, pp. 1285–1289.

[85] ——, "Enhanced Time of Arrival Estimation with Successive Cancellation," in *Proceedings IEEE Vehicular Technology Conference*, vol. 2, 2002, pp. 851–855.

[86] ——, "Performance of Successive Cancellation Techniques for Time of Arrival Estimation," in *Proceedings IEEE Vehicular Technology Conference*, vol. 4, 2002, pp. 2278–2282.

[87] C. Knapp and G. Carter, "The Generalized Correlation Method for Estimation of Time Delay," *IEEE Transactions on Acoustics, Speech, and Signal Processing*, vol. 24, no. 4, pp. 320–327, 1976.

[88] F.-X. Ge, D. Shen, Y. Peng, and V. O. K. Li, "Super-Resolution Time Delay Estimation in Multipath Environments," *IEEE Transactions on Circuits and Systems—Part I: Regular Papers*, vol. 54, no. 9, pp. 1977–1986, 2007.

[89] A.-J. van der Veen, M. C. Vanderveen, and A. J. Paulraj, "Joint Angle and Delay Estimation Using Shift-Invariance Properties," *IEEE Signal Processing Letters*, vol. 4, no. 5, pp. 142–145, 1997.

[90] A.-J. van der Veen, M. C. Vanderveen, and A. Paulraj, "Joint Angle and Delay Estimation Using Shift-Invariance Techniques," *IEEE Transactions on Signal Processing*, vol. 46, no. 2, pp. 405–418, 1998.

[91] M. Haardt and J. A. Nossek, "Unitary ESPRIT: How to Obtain Increased Estimation Accuracy with a Reduced Computational Burden," *IEEE Transactions on Signal Processing*, vol. 43, no. 5, pp. 1232–1242, 1995.

[92] M. C. Vanderveen, C. B. Papadias, and A. Paulraj, "Joint Angle and Delay Estimation (JADE) for Multipath Signals Arriving at an Antenna Array," *IEEE Communications Letters*, vol. 1, no. 1, pp. 12–14, 1997.

[93] B. H. Fleury, M. Tschudin, R. Heddergott, D. Dahlhaus, and K. Ingeman Pedersen, "Channel Parameter Estimation in Mobile Radio Environments Using the SAGE Algorithm," *IEEE Journal on Selected Areas in Communications*, vol. 17, no. 3, pp. 434–450, 1999.

[94] H. C. So, Y. T. Chan, Q. Ma, and P. C. Ching, "Comparison of Various Periodograms for Sinusoid Detection and Frequency Estimation," *IEEE Transactions on Aerospace and Electronic Systems*, vol. 35, no. 3, pp. 945–952, 1999.

[95] J. Capon, "High-Resolution Frequency-Wavenumber Spectrum Analysis," *Proceedings of the IEEE*, vol. 57, no. 8, pp. 1408–1418, 1969.

[96] C. Lombardi, M. Luise, and R. Reggiannini, "Channel Estimation and Equalization for Narrow-Band TDMA Mobile Radio," in *Proceedings IEEE International Conference on Communications*, 1990, pp. 1346–1350.

[97] Z. Ding and G. Li, "Single-Channel Blind Equalization for GSM Cellular Systems," *IEEE Journal on Selected Areas in Communications*, vol. 16, no. 8, pp. 1493–1505, 1998.

[98] D. Boss, K.-D. Kammeyer, and T. Petermann, "Is Blind Channel Estimation Feasible in Mobile Communication Systems? A Study Based on GSM," *IEEE Journal on Selected Areas in Communications*, vol. 16, no. 8, pp. 1479–1492, 1998.

[99] M. Morelli and U. Mengali, "Joint Frequency and Timing Recovery for MSK-Type Modulation," *IEEE Transactions on Communications*, vol. 47, no. 6, pp. 938–946, 1999.

[100] Y.-L. Huang, K.-D. Fan, and C.-C. Huang, "A Fully Digital Noncoherent and Coherent GMSK Receiver Architecture with Joint Symbol Timing Error and Frequency Offset Estimation," *IEEE Transactions on Vehicular Technology*, vol. 49, no. 3, pp. 863–874, 2000.

[101] J. Riba and G. Vazquez, "Parameter Estimation of Binary CPM Signals," in *Proceedings IEEE International Conference on Acoustics, Speech and Signal Processing*, vol. 4, 2001, pp. 2561–2564.

[102] U. Mengali and A. N. D'Andrea, *Synchronization Techniques for Digital Receivers*. Plenum Press, 1997.

www.ingramcontent.com/pod-product-compliance
Lightning Source LLC
Chambersburg PA
CBHW081516190326
41458CB00015B/5384